GO FIGURE!

The Numbers You Need for Everyday Life

GO FIGURE!

The Numbers You Need for Everyday Life

Nigel J. Hopkins • John W. Mayne • John R. Hudson

VISIBLE™

INK

PRESS

DETROIT   WASHINGTON, D.C.   LONDON

GO FIGURE!

Published by Visible Ink Press™
a division of Gale Research Inc.
835 Penobscot Building
Detroit MI 48226-4094

Visible Ink Press is a trademark of Gale Research Inc.

ISBN 0-8103-9424-3

Art Director: Arthur Chartow
Design Supervisor: Cindy Baldwin
Cover Design: Kathleen A. Mouzakis
Cover Illustration: Terry Colon

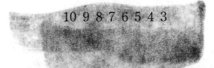

10 9 8 7 6 5 4 3

Contents

3 ◆ Bank on It..41

4 ◆ Health and Fitness by the Numbers..............65

5 ♦ Weather and the Environment........................97

6 ♦ Gambling, Cards, and Games— Odds and Probabilities 125

7 ♦ Sports—Scoring and Statistics 149

8 ♦ Home, Hobbies, and Workshop Numbers .171

9 ♦ Popular Science Calculations 209

Appendix A ♦ Tools You Might Need 241

Appendix B ◆ Getting the Units of Measurement Right 289

Appendix C ◆ More About Money 305

Suggestions for Further Reading 327

Index 333

Introduction

This book is not a refresher course in remedial math. It is for the majority of us who, with a few tools at our disposal, have the ability to get a better handle on some of life's everyday puzzlers. A large part of our well-being—financial and otherwise—depends on how well we understand the numbers we routinely encounter. But most of us don't have the time, patience, or paper to sit down and re-figure the hundreds of formulas we could use at work, home, and play to confidently make many day-to-day decisions or better understand the world around us.

How much are my credit cards costing me? Should I cancel some?

Is it time to sell my car?

What do those hockey stats mean?

When is the next Leap Year?

How much heat will I get out of this firewood?

How much wallpaper do I need?

What's my optimum weight?

How are my investments doing?

What are my utility bills telling me?

How can I convert this recipe to measurements I use in my kitchen?

How much should I tip at lunch?

What does the wind chill factor mean—do I need a warmer coat?

When a word confuses you, the remedy is simple: Grab a dictionary and look it up. But when a number or a calculation—such as those above—leaves you baffled, you probably don't have a handy reference to grab and consult.

A calculator can help you with routine arithmetic, but it can't tell you, for example, how much more a mortgage will cost if interest rates rise, or what your chances are of improving a poker hand if you draw two cards to three-of-a-kind, or what it means if your doctor says your blood pressure is 120 over 80.

For most of these answers, you'd have to consult a number of specialized sources. How much easier if you had a single, well-stocked toolkit full of formulas, tables, statistics, definitions, and examples to make everyday calculations more accessible. How much easier if, for instance, you could refer to a general handbook with answers to a wide range of numerical questions? Now you can!

Go Figure! is that book—a practical handbook that will guide you past numerical problems in many different areas, including: money; health and fitness; weather and the environment; gambling, cards, and games; sports and statistics; home and hobbies; and popular science.

In the appendices, you'll find helpful information that goes beyond the direct answers in the text. For example: Because units of measurement come into play in most numerical problems, often causing confusion, one appendix defines the units you might meet in everyday problems. It can help you if you encounter something like a coulomb or a kilopascal in your stroll through life.

We put this book together confident that it will prove so useful to you that you'll decide to keep a copy beside your dictionary, your cookbook, and other quick-reference tools in your home. You'll probably use it more often than you think. Go figure!

Acknowledgments

We are grateful for the assistance of our associates at Orbita Consulting Ltd.: W.P. Doyle, W.R. Waters, J. Laskoski, A.C. Jones, G.D. Kaye, and H.H. Watson. Each has worked on at least one section of the book, and their collective contributions have greatly enhanced its scope.

Through our suggestions for further reading, beginning on page 327, we acknowledge our debt to their authors and recommend them as sources of further information. We have also consulted publications of the following organizations: American Stock Exchange; Dow Jones Inc.; Franklin Associates Ltd.; International Amateur Athletic Federation; Moody's Investor Services; National Association of Securities Dealers; National Center for Health Statistics; New York Stock Exchange; Ontario Ministry of Natural Resources; Participaction; Standard and Poor's Corporation; State University of New Jersey; Uniform Code Council; U.S. Bureau of Economic Analysis; U.S. Bureau of the Census; U.S. Centers for Disease Control; U.S. Department of Agriculture; U.S. Department of Commerce; U.S. Department of Energy; U.S. Department of Labor; U.S. Department of Transportation; U.S. Environmental Protection Agency; U.S. Food and Drug Administration; U.S. Golf Association; U.S. Postal Service; U.S. Social Security Administration; Wilshire Corporation; and World Health Organization.

Three other contributors—Sally Carling, Jean Mayne, and Audrey Hudson—deserve special mention for reasons that they will understand.

Thank you to Patty Miller for reviewing the manuscript and offering useful comment.

Finally, we wish to thank our editors at Visible Ink Press and Gale Research Inc., notably senior editors Donald Boyden and Diane L. Dupuis, senior developmental editors Christine Hammes and Lawrence Baker, and developmental editor Rebecca Nelson. Their skills and encouragement have been invaluable.

Nigel J. Hopkins
John W. Mayne
John R. Hudson

About the Authors

Nigel Hopkins, John Mayne, and John Hudson are professionals at Orbita Consultants Ltd, specializing in operations research and systems analysis. Retired from the Canadian government, the three bring varied academic backgrounds to *Go Figure!* Hopkins's specialty is nuclear physics, Mayne's is mathematics, and Hudson's is computers.

1

Consuming Passion

The most important numbers in your life may be the numbers that concern your money. If you don't understand those numbers, the mistakes you make can get expensive.

This chapter covers common consumer money matters: information on credit cards and car-buying, how to figure your tip at a restaurant, foreign currency comparisons, and tips on understanding the Consumer Price Index. You can find more information on some of the topics in an appendix.

1.1—What Your Credit Cards Cost You

We pay for the convenience of using credit cards through interest charges and other fees. Different companies charge different rates, and they calculate the balance on which interest is charged in slightly different ways. The method described here is a representative one that will illustrate how such calculations can be made. For exact details, contact the company that issued your card.

The balance on which interest is charged is an average daily balance, calculated using the information provided in a typical statement:

TABLE 1.1A

Sample Statement

Date of Transaction	Posting Date	Amount
Oct. 15	Oct. 22	$ 10
Oct. 21	Oct. 28	$ 50
Nov. 7	Nov. 14	$ 50
Nov. 17	Payment	$210

Statement Date .... Nov. 25
Previous Statement Date .... Oct. 25
Previous Balance (including interest) $610
Previous Month Interest $10
Monthly Interest Rate 1.4375%

The method used in this example to calculate the average daily balance excludes interest payments and new purchases, but includes payments received and the previous balance (including the previous month's interest).

The billing cycle in this case is 31 days, running from October 25 to November 25.

Because interest is excluded, we subtract the interest from the previous balance and from the payment made, as follows:

Net previous balance: $610 – $10 = $600

Net payment made: $210 – $10 = $200

We note that the payment of $210 was made eight days prior to the statement date.

✏ Finding the Average Daily Balance

You may be interested in determining what the credit card company's average daily balance formula is.

The average daily balance is found as follows:

Step 1: Multiply the net previous balance by 31:
$600 × 31 = $18,600

Step 2: Multiply the net payment by 8:
$200 × 8 = $1,600

Step 3: Find the difference between the two figures:
$18,600 – $1,600 = $17,000

Step 4: Divide this figure by the number of days (31) in the billing cycle to find the average daily balance:
$17,000 ÷ 31 = $548.39

Step 5: To find the interest charge for the month, multiply this average daily balance by the monthly interest rate:
$548.39 × 1.4375% = $7.88

✏ Finding the Amount Owed

We can find the total amount owed for the month, as follows:

Previous balance due, including interest $610.00
plus current interest due ....$ 7.88
plus total of new purchases (10 + 50 + 50) $110.00
minus payment received $210.00

Total amount owed $517.88

You can see that, although you have paid $100 more than the total of your new purchases, the balance owed has been reduced by only $92.12 ($610 – $517.88) because part of your payment went toward interest charges.

⇨ Save Money by Paying Promptly

This example clearly indicates that it is always to your advantage to pay off all of the balance due each month in order to avoid paying high interest charges.

The monthly interest rate used in these calculations is a typical rate for one of the major credit-card companies. When multiplied by 12, it gives an annual rate of 17.25%—which is the annual rate that the lender quotes in its sales materials.

However, a better picture of the *true* interest rate you are paying is given by the annual effective rate, which is related to the quoted rate by the compound-interest formula in C.1, page 307.

In this case, the effective rate works out to 18.68%—which is higher than the quoted rate because it includes interest paid on interest as well as interest paid on principal.

In addition to interest charges, you may also have to pay either transaction fees (typically around 15 cents per transaction), or a fixed annual fee (typically about $20).

1.2—The Cost of Buying a New Car

Is it time for you to buy a new car? Good question.

Outlined here is a method of estimating the cost of a new car. If you have kept detailed records of the cost of owning your old car, use them in the proceeding calculations; if not, you can use estimates based on whatever data you can find for similar cars.

Cost Factors to Consider

The factors to consider include the following: trade-in value of your present car, the price of a new car, depreciation, insurance, repairs and maintenance, registration and licensing, and direct operating costs for fuel, oil, etc. In addition, you may need to consider the cost of borrowing money to finance the purchase and the investment income you will forgo because the money you spend on a new car will not be available for profitable investment. You will also need to take account of the additional trade-in value of a new car, as compared to the trade-in value of your old car.

In describing the method of comparison, it is convenient to use a set of representative figures and assumptions.

EXAMPLE

1: You own a three-year-old car that cost $10,000.

2: You have determined that a new car will cost $15,000.

3: Typical depreciation rates for both cars are 30% in the first year; 20% in the second year; 10% in the third and fourth years; and 5% in later years.

4: You have $6,000 in cash for the down payment, but using it for this purpose means your investment losses will amount to 10% per year, subject to tax at the rate of 31%.

Using the assumed depreciation rates, you calculate that the trade-in values of the cars will be as in Table 1.2A:

TABLE 1.2A

Trade-in Values of Cars

	OLD CAR	NEW CAR
Original value	$10,000	$15,000
After 1st year (–30%)	7,000	10,000
After 2nd year (–20%)	5,600	8,000
After 3rd year (–10%)	5,040	7,200
After 4th year (–10%)	4,540	6,840
After 5th year (–5%)	4,310	6,160
After 6th year (–5%)	4,090	5,850

From Table 1.2A, you see that your three-year-old car now has a trade-in value of about $5,040. (Check this estimate by calling several dealers to ask what they might offer and by looking at tables of used-car values, available in your local library.) Since the difference between the trade-in value of your old car and the new-car price is $10,000, and you have $6,000 cash, you determine that you will need to borrow $4,000. You check and find that you can borrow this amount at 12% interest, repayable over three years. From Table 2.3, page 28, you calculate that this will cost you $133 a month, or about $1,600 per year (12 × $133) for each of the three years of the loan.

Investment Losses

Using Table 3.1, page 47, you can determine that your losses (on $6,000 at 10%, with 31% taxes) would be $420 in the first year; $450 in the second year; $480 in the third; and $515 in the fourth year.

Of course, if you are still paying for your old car, you can leave its payments and investment losses out of your calculations because its costs will continue if you keep it.

Insurance Costs

Insurance costs vary with the value of a car. You can use your current costs for your old car, but you will need to make an estimate for the cost of insuring a new car. Some representative figures are used in this example: $500 per year for the old car and $700 per year for the new car.

➥ Direct Costs

You should base your estimates of direct operating costs—including repairs and maintenance—as much as possible on the costs you've actually experienced, since they reflect the type of driving you do, how many miles you drive, and so on. If you have not kept good records, ask your service center for estimates or consult tables of repair and maintenance costs that you'll find in your local public library. We assume here that direct operating costs for your present car are $700 per year and that you will not change your driving habits—but that a new car will get better mileage, so that your operating costs will be a little lower, at $650 per year.

➥ Comparative Costs of Options

Tables 1.2B and 1.2C summarize the values and costs for the two options: keeping your old car or buying a new one. The entries in each case are those that apply during the year listed. For example: During the first year after you buy a new car, its trade-in value is assumed to be its year-end value, and the costs for the year are assumed to be those you would pay if you kept the car for the full year. Since you have owned your present car for three years, the figures for it begin in the fourth year of its life. (The figures have been rounded off.)

TABLE 1.2B

Values and Costs for Your Old Car

YEAR	TRADE-IN VALUE	DEPREC-IATION	REPAIRS	OPER-ATING	LOAN COST	INVEST-MENT LOSSES	INSUR-ANCE	TOTAL ANNUAL COST
4th	$5,040	$500	$1,300	$700	0	0	$500	$3,000
5th	4,540	230	2,000	700	0	0	500	3,430
6th	4,310	220	2,700	700	0	0	500	4,120
7th	4,090	200	3,500	700	0	0	500	4,900

TABLE 1.2C

Values and Costs for a New Car

YEAR	TRADE-IN VALUE	DEPREC-IATION	REPAIRS	OPER-ATING	LOAN COST	INVEST-MENT LOSSES	INSUR-ANCE	TOTAL ANNUAL COST
1st	$10,000	$5,000	$650	$ 200	$1,600	$420	$700	$8,570
2nd	8,000	2,000	650	400	1,600	450	700	5,800
3rd	7,200	800	650	700	1,600	550	700	4,990
4th	6,840	360	650	1,300	0	600	700	3,610

✏ Potential Value of Your Assets

In this case, you can see that your annual cost will be considerably lower if you keep your old car for another three years—after which you will begin to make some savings from a new car. Note, however, that at the end of a new car's third year, you would have an asset with a potential value of $7,200 as compared to one worth $4,310— a difference of $2,890. At the end of the fourth year, there is still a difference of $2,750 in the trade-in values.

You may want to take this difference into account in your calculations, although the potential value of the difference will not be realized unless you sell your car.

Clearly, the initial depreciation on a new car is the biggest expense of a brand-new car. This expense can be avoided by buying a one or two-year-old car rather than a new car.

Later in the life of an older car, the cost of maintenance and repair generally becomes the dominant factor. But until that point is reached, it's cheaper to hold on to an older car.

1.3—Percentages: Markdowns, Markups, Taxes, and Tips

✏ **Subtracting a Percentage**

You will typically want to subtract a percentage when a price has been marked down and you want to find the new price.

A price of $35 is reduced by 15%, and you want to find the new price.

Step 1: Subtract the markdown percentage from 100%:
100% – 15% = 85%

Step 2: Multiply by the original price:
85% of $35 = .85 × $35 = $29.75

You don't have to be a math wizard, though, to figure percentages in your head: It's as easy as moving a decimal. Percentages may be figured by using ten percent as a base. For example, 10 percent of $35 is $3.50; so 5 percent of $35 will be half of that amount, or $1.75. Then you know that, by adding those figures, 15 percent of $35 is $5.25. $35 minus $5.25 equals $29.75. So you have the discounted price.

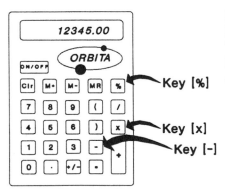

If you have a calculator with a % key, you can use it in this case. There are two ways of using the % key; you will have to check which one your calculator uses.

1: KEY[35], KEY[–], KEY[15], KEY[%]

2: KEY[35], KEY[×], KEY[15], KEY[%], KEY[–]

Figure 1.3A—Using a % Key for Price Cuts

✏ Adding a Percentage

You will typically meet this problem when the price of something is marked up by a percentage, or when a sales tax is added, or when you want to add a tip to your bill at a restaurant.

A price of $35 is to be increased by 15%, and you want to find the new price.

Step 1: Add the markup percentage to 100%:
100% + 15% = 115%

Step 2: Multiply by the original price:
115% of $35 = 1.15 × $35 = $40.25

Just as with markdowns, a tip may be figured using ten percent as the base. Ten percent of 35 is 3.50 (just move the decimal). We know then that 5 percent is 1.75 (3.50 ÷ 2). Adding ten percent (3.50) to five percent (1.75), you have 15 percent (5.25). The total bill is 35 + 5.25 = $40.25.

If you have a calculator with a % key, you can use it in this case. There are two ways of using the % key; you will have to check which one your calculator uses.

1: KEY[35], KEY[+], KEY[15], KEY[%]

2: KEY[35], KEY[×], KEY[15], KEY[%], KEY[+]

✏ Subtracting a Percentage and Adding Another

You will meet this problem, for instance, if a price is marked down by a percentage and a sales tax is added.

A price of $35 is marked down by 15%, and a sales tax of 5% is added. You want to find the new price.

Step 1: Subtract the markdown percentage from 100%:
100% – 15% = 85%

Step 2: Multiply by the original price:
85% of $35 = .85 × $35 = $29.75

Step 3: Add the sales-tax percentage to 100%:
100% + 5% = 105%

Step 4: Multiply by the reduced price:
105% of $29.75 = 1.05 × $29.75 =
$31.24

If you have a calculator with a % key, you can use it for this calculation, following the sequence that is correct for your machine, which will be one of the two following sequences:

1: KEY[35], KEY[–], KEY[15], KEY[%], KEY[+], KEY[5], KEY[%]

2: KEY[35], KEY[×], KEY[15], KEY[%], KEY[–], KEY[5], KEY[%], KEY[+]

Once again, referring to the "Subtracting a Percentage" section, you can use 10 percent as the base. After arriving at the markdown (in this case, 29.75), simply multiply by 1.05, which will add in the 5 percent sales tax and give you the final total of 31.24.

1.4—Exchanging Currencies

☞ Trading Your Dollars

Rates of exchange between different currencies vary from day to day. They are quoted monthly in the *Federal Reserve Bulletin* and averaged over the year on the basis of certified noon buying rates in New York for cable transfers. Table 1.4A shows average yearly rates for some of the world's main currencies. A complete listing of up-to-date values for many foreign currencies can usually be found in the financial sections of large newspapers.

TABLE 1.4A

Foreign Currency Units/U.S. Dollar

COUNTRY/UNIT	1989	1990	1991(1)
Australia/dollar	1.26	1.26	1.28
Austria/schilling	13.24	11.33	12.09
Belgium/franc	39.41	33.43	35.40
Canada/dollar	1.18	1.17	1.14
China/yuan	3.77	4.79	5.38
Denmark/krone	7.32	6.19	6.64
France/franc	6.38	5.45	5.85
Germany/deutsche mark	1.88	1.62	1.72
Hong Kong/dollar	7.80	7.79	7.75
India/rupee	16.21	17.49	24.99
Italy/lira	1372.3	1198.2	1284.7
Japan/yen	138.07	145.0	134.8
Netherlands/guilder	2.12	1.82	1.94
New Zealand/dollar	1.68	1.67	1.75
Norway/krone	6.91	6.25	7.22
Portugal/escudo	157.53	142.7	148.8
Spain/peseta	118.44	101.96	107.88
Sweden/krona	6.46	5.92	6.24
Switzerland/franc	1.63	1.39	1.50
Taiwan/dollar	26.41	26.92	26.64
United Kingdom/pound	.6104	.5605	.5874

(1) 6-month average, June to November

Source: Federal Reserve Bulletin (monthly)

If you plan on traveling in a country with unfamiliar currency, you may find it useful to prepare a card that you can keep in your pocket, giving the equivalent values of the coins and notes you will handle every day.

 If the current rate of exchange for the U.K. pound (£) is .600 per U.S. dollar, then $1 is worth £ .600, or 60 pence (P), and £ 1 is worth $1.666. Table 1.4B shows what you might include on your convenient pocket card.

TABLE 1.4B

U.S./U.K. Conversion chart

U.S.	U.K	U.K.	U.S.
$.05	3P	10P	$ .17
$.10	6P	20P	$ .33
$.25	15P	30P	$ .56
$.50	30P	50P	$ .83
$1	60P	£1	$ 1.67
$2	£1.20	£2	$ 3.33
$3	£1.80	£3	$ 5.00
$4	£2.40	£4	$ 6.67
$5	£3.00	£5	$ 8.33
$6	£3.60	£6	$10.00
$7	£4.20	£7	$11.67
$8	£4.80	£8	$13.33
$9	£5.40	£9	$15.00
$10	£6.00	£10	$16.67

1.5—Understanding the Consumer Price Index

The Consumer Price Index (CPI) is a measure of the relative level of prices for consumer goods and services. It is widely used as an indicator of the rate of inflation that the average consumer faces.

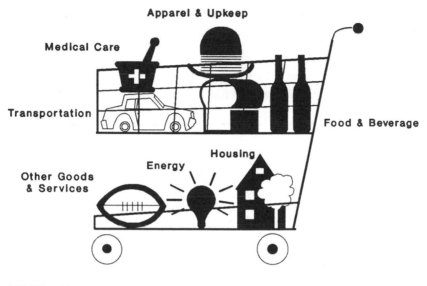

ORBITA 1991

Figure 1.5A—"Basket of Goods" in the CPI(U)

Symbolically, the CPI is depicted as a basket of goods that a typical consumer might bring to a cashier in a supermarket; in fact, however, it includes many items, such as housing and transportation, that are not grocery items. The CPI is not a true cost-of-living index because it does not include all items. Income taxes, for example, are not part of the basket.

☞ The CPI is a Relative Index

The CPI measures the net change, since a given base period, in the cost of the items in the basket of goods and services. The price of the CPI basket in the base period is assigned a value of 100, and the price at any later time is expressed as a percentage of the price in the base period. For example: If the price of the basket has increased 30% since the base year, then the index is 130; if the price has fallen 8%, the index is 92. It is important to understand that the CPI measures price increases or decreases, not actual price levels.

 You want to compare the prices of meat and bread.

If the index for meat is 110 and the index for bread is 120, it does not mean that bread is more expensive than meat. It means that the price of bread has increased by 20% since the base period, while the price of meat has increased by only 10% during the same period.

☞ Data are Collected Periodically

The information required to calculate the CPI is obtained from systematic surveys of actual consumer spending. Since 1978, the U.S. Labor Department has published two indexes: the CPI(U), covering about 80% of the population, including all urban consumers; and the CPI(W), covering about 32% of the population, including only urban wage earners and clerical workers. CPIs are currently published monthly and annually for many regions and selected cities in the U.S., as well as for a combined all-city average.

The periodic surveys also identify the items to be included in the "basket" and how much of the total spending goes toward each item. For example: Price changes are determined for each of a large number of food items, relative to their prices in the base period and their prices in the previous survey.

EXAMPLE Suppose that a loaf of bread cost 90 cents in the base period and $1.05 in the most recent survey, and you wish to compare each of their indexes.

Its index at the time of the most recent survey was: (1.05 × 100) ÷ 0.90 = 116.7. If the new survey finds that its price has increased to $1.10, its new index is: (1.10 × 100) ÷ 0.90 = 122.2.

The same result would be obtained by noting that the new index can be found by increasing the old index by the ratio of the two prices: that is, 116.7 × (1.10 ÷ 1.05) = 122.2.

Weights are Determined for the Categories

All of the items in a broad category, such as food, are grouped together, and a "weight" is determined from the percentage of total spending that goes toward buying all of the food items.

The major categories and the weight of each are typically as shown in Figure 1.5B:

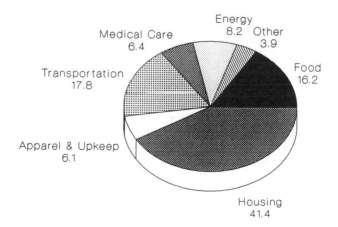

Figure 1.5B—CPI(U) Categories and Weights

Price indexes for various groups of items, as well as for the all-items CPI, are calculated by a process of aggregation which ensures that each item's index is assigned a weight, or relative importance, that reflects the actual spending pattern of the population to which the CPI refers.

The procedure is the same whether combining price indexes to form a sub-group index or combining the major component indexes of the CPI—i.e., those seven listed in Figure 1.5B. In Table 1.5A, the U.S. National CPI(U) is listed by category and by year. The base period now in use is the period 1982–84. All of the indexes are related to their assigned value of 100 in that period.

TABLE 1.5A

U.S. National CPI(U) by Category and by Year

	1984	1985	1986	1987	1988	1989	1990	1991*
ALL ITEMS.........	103.9	107.6	109.6	113.6	118.3	124.0	130.7	135.8
Food & beverages.........	103.2	105.6	109.0	113.5	118.2	125.1	132.4	136.3
Housing..............	103.6	107.7	110.9	114.2	118.5	123.0	128.5	133.2
Apparel & upkeep	102.1	105.0	105.9	110.6	115.4	118.6	124.1	128.5
Transportation	103.7	106.4	102.3	105.4	108.7	114.1	120.5	123.7
Medical care	106.8	113.5	122.0	130.1	138.6	149.3	162.8	176.1
Energy..............	100.9	101.6	88.2	88.6	89.3	94.3	102.1	102.0

* First 10 months
Source: U.S. Department of Labor

Table 1.5A can be used to compare one year's costs (1991's, for example) with those of another year (1984's, for instance) by multiplying the ratio of the indexes by the actual costs in the earlier year.

You spent $50 per week for food and beverages in 1984, and want to know the equivalent cost in 1991.

$50 is multiplied by the ratio of the Food & Beverage Indexes for 1991 and 1984:

$$\$50 \times 136.3 \div 103.2 = \$66.03.$$

This means that the food basket that cost $50 in 1984 would have cost $66.03 in 1991.

Figure 1.5C shows how the all-items CPI(U) has been steadily increasing over the years since 1982. It also shows how the scale has been set with the value 100 at an average for the years 1982-84.

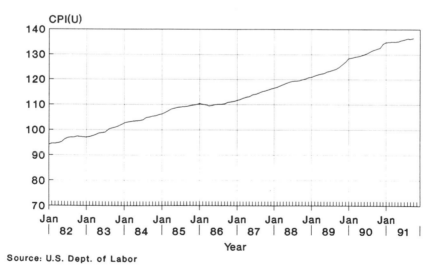

Source: U.S. Dept. of Labor

Figure 1.5C—CPI(U) Increases Since 1982

2

Your Money or Your Life

This chapter will fill you in on matters relating to investments, mortgages, IRAs—in short, money decisions that are usually considered long-term.

2.1—Reading the Financial Pages

More good questions:
- Can I afford to make some investments?
- If so, what sorts of investments?

In this section, we'll try to help you find the answers.

✎ The Percentage of Income Rule

Here is an old rule of thumb that can help you decide when it's appropriate to make some investments: Your savings should total about three to six months' income; beyond that, feel free to take some risks.

✎ The 100-year Investment Rule

Another useful rule of thumb concerns the type of investment you should consider: Subtract your age from 100, and ...

- If you are over 50, this number is the percentage of your funds to put into low- to moderate-risk investments;
- If you are under 50, the balance of 100 is the percentage of your funds you should put into growth-oriented, higher-risk investments;
- If you are around 50, you should spread your investments evenly: one-third in high-risk, one-third in moderate-risk, and one-third in low-risk investments.

✎ Bonds

Bonds issued by governments and corporations pay interest at a stated "coupon rate"—say, 9.25%—which are redeemable on the stated maturity date. When you multiply the coupon rate by $1,000 (the usual maturity price of a bond), the result is the annual interest you'll receive—usually in semiannual payments. (In this case: $1,000 × 9.25% = $92.50.) There are wide variations in the coupon rates of bonds, depending on maturity dates, convertibility into shares, market conditions, and the like. The financial pages list bonds and their yields.

⬯ Stocks

When you buy a share of stock in a company, you become a part-owner of that company. The value of that share fluctuates from day to day, as determined in an open marketplace of buyers and sellers.

Stock prices are reported daily on the financial pages. The listings usually show, from left to right: the range of prices for that stock over the past year; the stock name (using a code); the annual dividend (based on the most recent setting); the yield (the ratio of the dividend to the current price); the earnings per share (in the last fiscal year or the latest interim earnings); the price-to-earnings-per-share (P/E) ratio; the number of shares traded; yesterday's high, low, and closing prices; and the change in price.

⬯ Mutual Funds

Many investors choose mutual funds—a collection of stocks, bonds, or other securities purchased by a group of investors and managed by a professional investment company—because of the advantages of diversification and professional management. There are many such funds, with many different goals, and you will need to analyze them in terms of your own objectives. Periodically, financial magazines compare the performances of the various funds.

⬯ Stock Exchanges and Systems

The principal stock exchanges and stock systems that are reported on in the financial pages are: the New York Stock Exchange (NYSE); the American Stock Exchange (AMEX); and the National Association of Securities Dealers Automated Quotations (NASDAQ) National Market system.

The NYSE is the principal exchange for large-scale trading. About 1700 companies are listed.

The AMEX, also located in New York, is the main market for smaller companies and individuals. About 900 companies are listed.

The NASDAQ is the main market for over-the-counter (OTC) stocks (securities that are not listed on a stock exchange and are traded between individuals). About 4,300 companies are listed.

The Dow Jones Index

The Dow Jones Industrial Index is published daily as the principal indicator of the movement of stock prices. It represents an average of the combined values of the shares of 30 major corporations, including Exxon, General Electric, and McDonald's. Experience has shown that this index gives a good indication of the movement of all stocks being traded. It is reported very widely, and the leading financial papers plot graphs of its values over the past six months and the past year.

Other Indexes

There are three other Dow Jones Indexes: the Transportation Index, covering 20 companies; the Utilities Index, covering 15 companies; and the Composite Index, covering all the companies included in the Industrial, Transportation, and Utilities indexes.

The Standard and Poor's 500 Index weights its stocks according to its market value. The "500" consist of 400 industrial, 40 financial, 40 utility, and 20 transportation companies.

The Wilshire 5000 Index covers 5000 common stocks, including NYSE, AMEX, and OTC stocks.

The NASDAQ Composite Index covers all the stocks traded on its exchange.

2.2—Annual Equivalent Interest Rate on Bonds and Treasury Bills

You can often buy bonds and other types of debt instruments at a discount—and you may want to calculate the *actual* interest rate you will earn.

 You pay $961 for a Treasury bill maturing at $1,000 in 91 days, and you want to know what the interest rate is.

Since $961 spent now will be worth $1,000 after 91 days, the interest factor is 1000 ÷ 961 = 1.0405. (See C.1, page 307.) This interest is earned in one period of 91 days—and, therefore, the interest rate per period is 4.05%. Since there are four such periods in a year, the annual equivalent rate is 4 × 4.05 = 16.2% (See Table C.1, page 310.)

If taxes apply, the equivalent annual rate must be reduced by the applicable tax rate.

 If you must pay a marginal tax rate of 28%, the equivalent annual rate in the previous case would be: 16.2 − (28% of 16.2) = 11.66%.

2.3—The Monthly Payments on Your Short Term Loan

Most consumer loans—taken out for the purchase of cars, appliances, and other personal goods—are relatively short-term loans; they must be paid off over five years or less. You can calculate the monthly payments on these loans using Table 1.10, but it is more convenient to use Table 2.3, which shows entries covering some of the most typical cases. (You'll find the mathematical basis of this table in C.6, page 322.)

Table 2.3 shows monthly payments for payoff periods of two years, three years, four years and five years; for interest rates of 10% through 17% (compounded annually); and for loan amounts of $4,000 through $14,000. The results can be extended to larger and smaller amounts.

 EXAMPLE

To calculate monthly payments on a loan of $1,000, divide by 10 the results for $10,000; to calculate monthly payments on a loan of $16,000, multiply by two the results for $8,000; and so on.

You can calculate intermediate results by interpolating between values in the table using the formula given in A.2, page 247.

Table 2.3 also shows the total approximate interest charges that will be added to the principal over the period of the loan.

 EXAMPLE

The monthly payments on a five-year loan of $4,000 at 12% interest are $89, and you want to find out how much interest you will pay.

According to Table 2.3, the total interest charges are found to be $1,339—which means that by the time the loan has been paid off, the total amount repaid will be $4,000 + $1,339 = $5,339.

TABLE 2.3

Monthly Payments and Interest Charges on a Short-term Loan

AN-NUAL RATE	LOAN AMOUNT	2 YEARS M'THLY P'MENT	2 YEARS TOTAL INT.	3 YEARS M'THLY P'MENT	3 YEARS TOTAL INT.	4 YEARS M'THLY P'MENT	4 YEARS TOTAL INT.	5 YEARS M'THLY P'MENT	5 YEARS TOTAL INT.
10%	$ 4,000	$185	$430	$129	$647	$101	$870	$ 85	$1,099
	6,000	277	645	194	970	152	1,305	127	1,649
	8,000	369	860	258	1,293	203	1,739	170	2,199
	10,000	461	1,075	323	1,616	254	2,174	212	2,748
	12,000	554	1,290	387	1,939	304	2,689	255	3,298
	14,000	646	1,505	452	2,263	355	3,044	297	3,847
11%	$ 4,000	$186	$474	$131	$715	$103	$963	$87	$1,218
	6,000	280	712	196	1,072	155	1,444	130	1,827
	8,000	373	949	262	1,429	207	1,925	174	2,436
	10,000	466	1,186	327	1,786	258	2,406	217	3,045
	12,000	559	1,423	393	2,143	310	2,887	261	3,655
	14,000	653	1,660	458	2,500	362	3,368	304	4,264
12%	$ 4,000	$188	$519	$133	$783	$105	$1,056	$89	$1,339
	6,000	282	779	199	1,174	158	1,584	133	2,008
	8,000	377	1,038	266	1,566	211	2,112	178	2,677
	10,000	471	1,298	332	1,957	263	2,640	222	3,347
	12,000	565	1,557	399	2,349	316	3,168	267	4,016
	14,000	659	1,817	465	2,740	369	3,696	311	4,685
13%	$ 4,000	$190	$564	$135	$852	$107	$1,151	$91	$1,463
	6,000	285	846	202	1,278	161	1,727	137	2,194
	8,000	380	1,128	270	1,704	215	2,302	182	2,925
	10,000	476	1,415	337	2,135	268	2,882	228	3,657
	12,000	571	1,698	404	2,562	322	3,459	273	4,388

AN-NUAL RATE	LOAN AMOUNT	2 YEARS		3 YEARS		4 YEARS		5 YEARS	
		M'THLY P'MENT	TOTAL INT.	M'THLY P'MENT	TOTAL INT.	M'THLY P'MENT	TOTAL INT.	M'THLY P'MENT	TO-TAL INT.
	14,000	666	1,981	472	2,989	376	4,035	319	5,119
14%	$ 4,000	$192	$609	$137	$922	$109	$1,247	$93	$1,584
	6,000	288	914	205	1,383	164	1,870	140	2,376
	8,000	384	1,218	273	1,843	219	2,494	186	3,169
	10,000	480	1,523	342	2,304	273	3,117	233	3,961
	12,000	576	1,828	410	2,765	328	3,748	279	4,753
	14,000	672	2,132	478	3,225	383	4,363	326	5,545
15%	$ 4,000	$194	$655	$139	$992	$111	$1,344	$95	$1,710
	6,000	291	982	208	1,488	167	2,015	143	2,565
	8,000	388	1,310	277	1,984	223	2,687	190	3,420
	10,000	485	1,638	347	2,480	278	3,360	238	4,275
	12,000	582	1,965	416	2,977	334	4,032	285	5,130
	14,000	679	2,293	485	3,473	390	4,704	333	5,985
16%	$ 4,000	$196	$700	$141	$1,063	$113	$1,441	$97	$1,838
	6,000	294	1,051	211	1,594	170	2,162	146	2,757
	8,000	392	1,401	281	2,126	227	2,882	195	3,675
	10,000	490	1,755	352	2,660	283	3,607	243	4,594
	12,000	588	2,106	422	3,192	340	4,328	292	5,513
	14,000	686	2,457	492	3,724	397	5,050	341	6,432
17%	$ 4,000	$198	$746	$143	$1,134	$115	$1,540	$99	$1,965
	6,000	297	1,120	214	1,701	173	2,310	149	2,947
	8,000	396	1,493	285	2,268	231	3,080	199	3,929
	10,000	494	1,866	357	2,835	289	3,850	249	4,911
	12,000	593	2,239	428	3,402	346	4,620	298	5,894
	14,000	692	2,612	499	3,969	404	5,390	348	6,876

2.4—Your Monthly Mortgage Payments and Mortgage "Points"

When monthly mortgage payments are calculated, the amount of the payment that accounts for interest is blended with the amount owed on the principal, so that the total payment is the same for each month over the life of the mortgage. This means: In the early stages of the mortgage period, nearly all of the payment is interest; in the later stages, nearly all is principal. Use Table 2.4 to find the monthly payment, including principal and interest, that you will owe on each $1,000 of a mortgage. (You'll find the mathematical basis of the table in C.6, page 322.)

For intermediate interest rates, you will have to interpolate between the values in the table. (See A.2, page 247.)

 You want to calculate the payment on a mortgage of $60,000 amortized over 30 years at 11%, compounded semi-annually (as is typical for mortgage interest).

Table 2.4 shows $9.34 for the monthly payment on a $1,000 mortgage. Consequently, for a $60,000 mortgage, the payments will be 60 × 9.34 = $560.40 per month.

Sometimes mortgages are taken out on a variable-rate and variable-payment basis. Using Table 2.4, you can find the change in the payment if the interest rate varies.

 On a 20-year, $1,000 mortgage at 10%, compounded semi-annually, the payment is $9.52 per month. This will increase to $10.80 per month if the interest rate rises to 12%, and decrease to $8.89 per month if the interest rate falls to 9%.

The calculations in this section assume semi-annual compounding, since that is the common practice for mortgages paid monthly. If your lender uses a different compounding period, the effects are minimal—and can be ignored for approximate calculations.

People sometimes take out second mortgages on which the monthly payments cover the interest only. In that case, the approximate payment can be calculated as 1/12 of the annual interest.

 EXAMPLE The monthly payment on an $11,000 second mortgage at 12% will be 1% of $11,000, or $110.

In other cases, second mortgages require that part of the principal be paid off over a fixed period and that the remaining principal be due in a "balloon" payment at the end of that period. In such cases, the calculations can be made using Table 2.4 and other methods.

APR and Points

You may hear the terms "annual percentage rate" (APR) and "points" for a mortgage. A point is 1% of the amount of the mortgage. You may pay this when you take out the mortgage, but usually it is added to the principal, along with such other expenses as the appraisal fee. The interest calculated on this higher principal is the APR. You should be told both interest rates when you are taking out a mortgage. Sometimes, a seller will pay the points as a way to reduce the interest rate you'll have to pay—thus making the deal more attractive for you. (By law, a buyer may pay no more than 1 point on a VA or FHA mortgage.)

TABLE 2.4

Monthly Payments on a $1,000 Mortgage Amortized Over a Period of Years at Interest Rates Compounded Semi-annually (Dollars/month)

YRS	8%	9%	10%	11%	12%	13%	14%	15%	16%	17%	18%
15	9.48	10.0	10.6	11.2	11.8	12.4	13.1	13.7	14.3	15.0	15.6
16	9.17	9.75	10.3	10.9	11.5	12.2	12.8	13.5	14.1	14.8	15.4
17	8.90	9.59	10.1	10.7	11.3	12.0	12.6	13.3	13.9	14.6	15.3
18	8.67	9.26	9.87	10.5	11.1	11.8	12.4	13.1	13.8	14.5	15.1
19	8.46	9.07	9.68	10.3	11.0	11.6	12.3	13.0	13.6	14.3	15.0
20	8.28	8.89	9.52	10.1	10.8	11.5	12.2	12.8	13.5	14.2	14.9
21	8.12	8.74	9.37	10.0	10.7	11.4	12.0	12.8	13.4	14.1	14.9
22	7.97	8.60	9.25	9.90	10.6	11.3	11.9	12.7	13.4	14.1	14.8
23	7.85	8.48	9.13	9.80	10.5	11.2	11.8	12.6	13.3	14.0	14.7
24	7.73	8.38	9.03	9.71	10.4	11.1	11.7	12.5	13.2	14.0	14.7
25	7.63	8.28	8.95	9.63	10.3	11.0	11.7	12.5	13.2	13.9	14.7
26	7.54	8.19	8.87	9.55	10.3	11.0	11.7	12.4	13.1	13.9	14.6
27	7.45	8.11	8.79	9.49	10.2	10.9	11.6	12.4	13.1	13.9	14.6
28	7.38	8.05	8.73	9.43	10.1	10.9	11.6	12.3	13.0	13.8	14.6
29	7.31	7.98	8.67	9.38	10.1	10.8	11.6	12.3	13.1	13.8	14.6
30	7.25	7.93	8.63	9.34	10.1	10.8	11.5	12.3	13.0	13.8	14.5

2.5—Benefits from Your Individual Retirement Account (IRA)

Contributing to an IRA can significantly lower your tax bill if the contributions themselves are tax-free when you make them. The Internal Revenue Service (IRS) allows you to deduct up to $2,000 from your income tax depending on your income and your filing status. If, however, you were covered by a retirement plan through your employer, your IRA deduction may be reduced or eliminated. Of course the interest you earn on the account is taxable only when you withdraw it. Consult Table 2.5 to see the advantages of such IRA contributions over deposits in a regular savings account. (You'll find the mathematical basis for this table in C.6, page 322.)

The upper part of Table 2.5 shows the balance in an IRA if you deposit $100 at various interest rates. The table assumes that you'll pay no tax on the deposit and that each year's interest will be added to the principal at the end of each year. The lower parts of the table show the balances in a regular savings account if the $100 is taxable before it is deposited and the interest earned each year is taxable in that year.

 You have $100 to invest and want to choose between an IRA and a regular savings account.

If you contribute $100 to an IRA at 9% interest, Table 2.5 shows that after 15-years it would amount to $385. At a 28% tax rate, withdrawing this amount would leave you with $277.20.

If you save $100 in a regular account, it would first be taxed—at 28% in this example, so that only $72 would be deposited. The interest earned each year would be reduced by the same 28% tax, and the balance after 15 years would be $192. You'd owe no further tax on the regular account balance, leaving you with $192.

Best deal? The IRA would be worth 44% more than the regular savings account ($277.20 versus $192).

You can also use Table 2.5 to find the balance built up in an IRA by a series of yearly contributions. Do so by treating each contribution separately and adding the results.

You want to contribute $100 a year for five years and you want to know how much you'll have after five years.

At 9%, Table 2.5 shows that at the end of five years, the first contribution will have grown to $157; the second contribution, after four years, will have amounted to $143; and so on.

The combined total will be:

$$157 + 143 + 131 + 120 + 109 = \$660.$$

By contrast, the balance in a regular account at a tax rate of 28% would be:

$$100 + 94 + 88 + 82 + 77 = \$441.$$

As explained in Section 3.1, pages 44 to 46, you can make adjustments to take account of the following changes:

- Amounts other than $100
- Interpolating between values
- Different compounding periods
- Effects of inflation

TABLE 2.5

Future Value of $100 in an IRA Compared to Its Value in a Regular Savings Account (Dollars)

TAX RATE	YRS	INTEREST RATE										
		3%	4%	5%	6%	7%	8%	9%	10%	11%	12%	15%
	1	103	104	105	106	107	108	109	111	112	113	116
	2	106	108	111	113	115	117	120	122	125	127	135
IRA.	3	109	113	116	120	123	127	131	135	139	143	157
	4	113	117	122	127	132	138	143	149	155	162	182
	5	116	122	128	135	142	149	157	165	173	182	212
(NO	6	120	127	135	143	152	162	171	182	193	205	246
	7	123	132	142	152	163	175	188	201	216	232	286
TAX)	8	127	138	145	162	175	196	205	222	241	261	332
	9	131	143	157	172	188	205	225	246	269	294	386
	10	135	149	165	182	201	222	246	272	300	332	448
	12	143	161	182	205	232	261	294	332	374	422	605
	15	157	182	212	246	286	332	385	448	520	604	948
15%	1	87	88	89	89	90	91	92	93	93	94	97
	2	89	91	93	94	96	97	99	101	103	104	110
	3	92	94	97	99	102	104	107	110	113	116	125
	4	94	97	101	104	108	112	116	120	124	128	142
	5	97	101	105	110	115	120	125	130	136	142	162
	6	99	104	110	116	122	128	135	142	150	158	184
	7	102	112	115	122	129	137	146	155	164	175	209
	8	104	112	120	128	137	147	157	169	181	193	238
	9	107	115	125	135	146	157	170	184	198	214	271
	10	110	119	130	142	155	168	183	200	218	238	308
	12	115	128	142	157	174	193	214	237	263	292	399
	15	125	142	161	183	208	237	269	307	349	397	587
28%	1	74	74	75	75	76	76	77	77	78	79	80

TAX RATE	YRS	INTEREST RATE										
		3%	4%	5%	6%	7%	8%	9%	10%	11%	12%	15%
	2	75	76	77	79	80	81	82	83	85	86	90
	3	77	79	80	82	84	86	88	90	92	94	100
	4	79	81	83	86	88	91	94	96	99	102	112
	5	80	83	86	90	93	96	100	104	108	112	125
	6	82	86	89	93	98	102	107	112	117	122	139
	7	84	88	93	98	103	108	114	120	126	133	156
	8	86	91	96	102	108	115	122	129	137	145	174
	9	87	93	100	107	114	122	130	139	148	159	194
	10	89	86	103	111	120	129	139	149	161	173	217
	12	93	102	111	121	133	145	158	173	189	206	270
	15	109	111	124	138	154	172	192	215	240	269	376
31%	1	70	71	71	72	72	73	75	74	75	75	77
	2	72	73	74	75	76	77	78	79	81	82	85
	3	73	75	77	78	80	82	83	85	87	89	95
	4	75	77	79	82	84	86	89	91	94	97	105
	5	77	79	82	85	88	91	94	98	101	105	117
	6	78	81	85	89	92	96	101	105	110	114	130
	7	80	84	88	92	97	102	107	113	118	124	145
	8	81	86	91	96	102	108	114	121	128	135	161
	9	83	89	94	101	107	114	122	130	138	147	179
	10	85	91	98	105	112	121	129	139	149	160	199
	12	89	96	105	114	124	135	147	160	174	190	246
	15	94	105	116	129	143	159	177	197	219	244	337

2.6—How Much Life Insurance Do You Need?

The primary purpose of life insurance is to ensure that a surviving beneficiary will have an income sufficient to meet his or her needs. Consult Table 2.6 to see how much capital is needed to provide a certain monthly income over various numbers of years. (You'll find the mathematical basis for this table in C.6, page 323.) Table 2.6 assumes the invested capital will earn interest and that taxes will be owed on the income (which will be made up of interest and a portion of the original capital).

 Your beneficiary will need an after-tax income of $500 per month over a period of 20 years and you want to know how much insurance to buy.

If your survivor's tax rate will be 28%, and the proceeds of the policy can be invested to yield 9%, compounded daily, from Table 2.6, you find that you'll need $133 of principal for each $1 per month of after-tax income. Consequently, you will need 500 × $133 = $66,500 of insurance.

Note that the entries in Table 2.6 correspond to those in Table 3.5, page 59, because they also show the present value of a series of monthly payments to be made to the beneficiary. For convenience, Table 2.6 extends the period of years to 30 in order to simplify calculating payments that continue for many years.

As explained in Section 3.1, pages 44 to 46, you can make adjustments to take account of the following changes:
- Amounts other than $1 per month
- Interpolating between values
- Different compounding periods
- Effects of inflation

TABLE 2.6

Insurance Needed for an After-tax Income of $1 per Month for the Years Indicated While the Proceeds Earn Daily Interest (Dollars)

TAX RATE	YRS	3%	4%	5%	6%	7%	8%	9%	10%	11%	12%	15%
0%	2	23	23	23	23	22	22	22	22	21	21	21
	4	45	44	43	43	42	41	40	39	39	38	36
	6	66	64	62	60	59	57	55	54	52	51	47
	8	85	82	79	76	73	71	68	66	64	61	56
	10	104	99	94	90	86	82	79	76	72	70	62
	12	121	114	108	102	97	92	88	84	80	76	66
	14	137	128	121	113	107	101	95	90	85	81	70
	16	152	142	132	123	115	108	101	95	90	85	72
	18	167	154	142	132	122	114	107	100	94	88	74
	20	180	165	151	139	129	119	111	103	97	91	76
	25	211	185	171	155	141	129	119	110	102	95	78
	30	237	209	186	167	150	136	124	114	105	97	79
15%	2	23	23	23	23	23	22	22	22	22	22	21
	4	46	45	44	43	43	42	41	41	40	39	37
	6	67	65	63	62	60	59	57	56	55	53	50
	8	87	84	81	79	76	74	71	69	67	65	60
	10	106	102	98	94	90	87	83	80	77	75	67
	12	124	118	113	107	103	98	94	90	86	82	73
	14	141	133	126	120	114	108	103	98	93	89	77
	16	157	148	139	131	123	116	110	104	99	94	81
	18	173	161	151	141	132	124	117	110	104	98	84
	20	188	174	161	154	140	130	122	115	108	101	86
	25	222	202	184	169	155	143	133	123	115	107	89
	30	251	225	203	184	167	152	140	129	119	111	91
28%	2	23	23	23	23	23	23	22	22	22	22	22
	4	46	45	45	44	43	43	42	41	41	40	39

TAX RATE	YRS	3%	4%	5%	6%	7%	8%	9%	10%	11%	12%	15%
						INTEREST	RATE					
	6	67	66	65	63	62	61	59	58	57	56	51
	8	88	86	83	81	79	77	74	73	71	69	63
	10	108	104	101	97	94	91	88	85	85	90	72
	12	127	122	117	112	108	103	99	96	92	89	79
	14	145	138	132	126	120	115	110	105	101	96	85
	16	162	154	145	138	131	125	119	113	108	103	90
	18	179	168	158	149	141	134	126	120	114	108	93
	20	195	182	117	160	156	141	133	126	119	113	96
	25	231	213	197	182	169	158	147	138	129	122	101
	30	265	240	219	201	184	170	157	146	135	126	104
31%	2	23	23	23	23	23	23	22	22	22	22	22
	4	46	45	45	44	44	43	42	42	42	41	39
	6	68	66	65	64	62	61	60	59	59	56	53
	8	88	86	84	81	79	77	75	73	73	70	64
	10	108	105	101	98	95	92	89	86	86	81	74
	12	127	122	118	113	109	105	101	97	97	90	81
	14	148	139	133	127	121	116	111	107	102	98	87
	16	163	155	147	148	133	127	121	115	110	105	92
	18	180	170	160	152	143	136	129	122	116	111	96
	20	196	184	173	162	153	114	136	129	122	116	99
	25	234	216	200	186	173	161	151	141	133	125	105
	30	268	244	233	205	189	174	161	150	140	121	108

3

Bank on It

One of the most common tasks that you'll perform is banking. Whether it's every few days when stopping at an automatic bank teller (ATM) to withdraw cash, every couple weeks when writing those checks to cover your bills, or every month when storing away a "little something," banking is a familiar part of your life. This chapter looks at interest on savings accounts, shows you how to deposit to reach a predetermined goal, and more.

3.1—Interest on a Single Deposit in Your Savings Account

⮕ Take an Interest in the Numbers

Use Table 3.1 to find the interest you would earn on $100 deposited in a daily-interest savings account and left on deposit for a number of years. (For the mathematical basis of this table, see C.6, page 323).

Use the interest rate to select the appropriate column, and use the number of years to select the appropriate row. The upper part of the table shows the results when the interest you earn is tax-free.

 You have deposited $100 at 8% (compounded daily), and you want to find the interest earned after 10 years.

In the column for 8% and the row for 10 years, you find the answer: $122.

⮕ Paying Taxes on Your Interest

You can use the lower parts of Table 3.1 when you are paying taxes on the interest earned but are leaving the rest of the interest on deposit in the account. The tax rates in the table are the current rates of federal income tax. (See C.3, page 318, for a discussion of the effects of taxes and the marginal tax rate.)

 You deposit $100 at 8% (compounded daily). Your marginal tax rate is 28%.

To find how much interest you will have netted after 10 years, you look in the column for 8% and the row for 10 years in the "28%" part of the table and find the answer: $79.

⮕ Finding the Balance

To find the balance in your account at the end of any year, you should add to the original principal the interest earned up to that time.

☞ Amounts Other than $100

To find answers for amounts other than $100 using Table 3.1, you only need to find the value for $100 and multiply by the appropriate factor. For instance: For $200, the table value would be multiplied by 2; for $350, it would be multiplied by 3.5; for $75, it would be multiplied by .75; and so on.

 EXAMPLE **The interest earned on $75, after 15 years at 7%, is found by multiplying the value $186 from Table 3.1 by 0.75 (75 ÷ 100). The answer: $139.50.**

☞ Interpolating Between Values in the Table

If the interest rate is between two of the values in the table, you will have to interpolate between the entries for the next-lower value and the next-higher value. (See A.2, page 247, to find a general formula for interpolating between values.)

 EXAMPLE **To find the interest earned after 10 years at a daily compounded rate of 7.5%, you must interpolate between the entries for 7% and 8% in Table 1.2.**

Using the "0%" portion of the table, you find the entry for 7% is 101 and for 8% is 122. Putting these in the formula gives:

$$I = (122 - 101) \times \frac{(7.5 - 7)}{(8 - 7)} = 21 \times .5 = 10.5$$

Therefore, the value for 7.5% interest is:

101 + 10.5 = 111.5

⇨ Extending the Number of Years

To find the interest earned over longer periods than the table covers—for example, for 20 years—you would first find the balance for 15 years. Use that number as the new principal, and find the interest on that principal for a further 5 years to obtain the answer for the full 20 years.

⇨ Different Compounding Periods

Table 3.1 gives results for cases in which interest is compounded daily. To find results for annual, semi-annual, quarterly or monthly compounding, you must determine the equivalent daily-interest rate by the method explained in C.4, page 319. With this equivalent interest rate, you can use Table 3.1 as before.

 To find the interest earned after 10 years on $100 at an interest rate of 9% compounded quarterly (assuming that no taxes are paid), Table C.4, page 319, shows that the equivalent daily interest rate is 8.9%.

From Table 3.1, the amount for 8% is $122 and the amount for 9% is $146. By the interpolation formula (page 247), the amount for 8.9% is:

$$122 + [0.9 \times (146-122)] = \$144$$

⇨ Effects of Inflation

Thus far, we have ignored the effects of inflation. When inflation is taken into account, you can estimate the relative buying power of the future balance in your savings account. (See C.5, page 321, for a discussion of the effects of inflation.)

 To find the interest earned after 10 years on $100 at an interest rate of 9% compounded daily, Table 3.1 shows the entry $146.

Added to the initial principal of $100, this makes a balance of $246. If the annual inflation over the intervening decade is forecast to be 6%, then the relative buying power of the balance at the end of 10 years will be found by looking up the entry 56 in Table C.5, page 321, and multiplying by 2.46 (246 ÷ 100) to obtain the value $138. This means that the effective interest rate over the period has been about 3% (the difference between the savings rate and the inflation rate). If you had paid taxes on the interest over the years, the buying power of the balance would have been much less than the present buying power of the principal.

✎ Some Useful Estimating Rules

A well-known way to estimate the growth of savings that are earning compound interest is the "Rule of 72," which says that the number of years needed to double an initial amount is found by dividing 72 by the annual interest rate.

 In Table 3.1, the original principal has doubled when interest earned is $100. You can see that the value 100 appears, as the rule indicates, at 10 years for 7%, at 7 years for 10%, and so on.

Table 3.1 illustrates that the Rule of 72 works. It also shows that similar rules can be found that can be useful in other cases. By studying Table 3.1, you can see that there are similar rules for cases in which taxes are being paid on the interest. The factors corresponding to the 72 in the original rule are as follows:

    82 for a 15% tax rate;

    95 for a 28% tax rate;

    99 for a 31% tax rate.

TABLE 3.1

Interest Earned, After Taxes, on a Single Deposit
of $100 in a Daily-interest Savings Account
(Dollars)

TAX RATE	YRS.	3%	4%	5%	6%	7%	8%	9%	10%	11%	12%	15%
						INTEREST	RATE					
0%	1	3	4	5	6	7	8	9	11	12	13	16
	2	6	8	11	13	15	17	20	22	25	27	35
	3	9	13	16	20	23	27	31	35	39	43	57
	4	13	17	22	27	32	38	43	49	55	62	82
	5	16	22	28	35	42	49	57	65	73	82	112
	6	20	27	35	43	52	62	71	82	93	105	146
	7	23	32	42	52	63	75	88	101	116	132	186
	8	27	38	49	62	75	90	105	122	146	161	232
	9	31	43	57	72	88	105	125	146	169	194	286
	10	35	49	65	82	101	122	146	172	200	232	505
	12	43	61	82	105	132	161	194	232	274	322	505
	15	57	82	112	146	186	232	285	348	420	584	848
15%	1	3	3	4	5	6	7	8	9	10	11	14
	2	5	7	9	11	13	15	17	19	21	23	29
	3	8	11	14	17	20	23	26	29	33	36	47
	4	11	15	19	23	27	31	36	41	46	51	67
	5	14	19	24	29	35	41	47	53	60	67	90
	6	17	23	29	36	43	51	59	62	76	85	117
	7	20	27	35	43	52	61	71	82	93	105	146
	8	23	31	41	51	61	73	85	98	112	128	180
	9	26	36	47	59	71	85	100	116	133	152	219
	10	29	41	53	67	82	98	116	135	156	180	263
	12	36	50	67	85	105	137	152	179	210	243	369
	15	47	67	90	116	145	179	217	261	311	368	591
28%	1	2	3	4	5	6	7	8	8	9	9	12

TAX RATE	YRS.	3%	4%	5%	6%	INTEREST RATE 7%	8%	9%	10%	11%	12%	15%
	2	4	6	8	9	11	12	14	16	17	19	25
	3	7	9	11	14	16	19	22	24	27	30	39
	4	9	12	16	19	23	26	30	34	38	42	55
	5	11	16	20	24	29	34	39	44	49	55	73
	6	14	19	24	30	36	42	48	55	62	69	94
	7	16	22	29	36	43	50	58	67	75	85	116
	8	19	26	34	42	50	59	69	79	90	102	141
	9	22	30	39	48	58	69	80	93	106	120	170
	10	24	34	44	55	66	79	93	107	123	141	201
	12	30	41	54	69	84	101	120	140	162	187	275
	15	38	54	72	92	114	139	167	199	234	273	422
31%	1	2	3	4	4	5	6	6	7	8	9	11
	2	4	6	7	9	10	12	13	15	17	18	24
	3	6	9	11	13	16	18	21	23	26	29	37
	4	9	12	15	18	22	25	29	32	36	40	53
	5	11	15	19	23	28	32	37	42	47	52	70
	6	13	18	23	28	34	40	46	52	59	66	89
	7	16	21	28	34	46	48	55	63	72	80	110
	8	18	25	32	40	48	56	65	75	85	96	133
	9	21	28	37	46	55	65	76	88	100	113	159
	10	23	32	42	52	63	75	88	101	116	132	188
	12	28	39	52	65	80	95	113	132	152	175	256
	15	37	52	68	87	108	131	157	186	218	254	389

3.2—Single Deposit You Need to Reach a Future Goal

The amount you must deposit today to reach a certain balance at some time in the future can be found using Table 3.2. (You'll find the mathematical basis for this table in C.6, page 323; as shown by the definition of "present value" in C.2, page 316. Table 3.2 also gives the present values of future payments.)

Table 3.2 has been calculated for a daily-interest account; it assumes that taxes are paid at various rates out of the interest at the end of each year. The table shows the amount you must deposit to achieve a balance of $100 at various years in the future.

 EXAMPLE **To accumulate a balance of $100 at the end of five years with a daily-interest rate of 6% and a marginal tax rate of 28%, Table 3.2 shows that you must deposit $80.**

As explained in Section 3.1, pages 44 to 46, you can make adjustments to take account of the following changes:
- Amounts other than $100.
- Interpolating between values.
- Different compounding periods.
- Effects of inflation.

TABLE 3.2

Amount of a Single Deposit Needed to Accumulate a Future Balance, After Taxes, of
$100 with Daily Interest Added (Dollars) (Present
Value of a Future Payment)

TAX RATE	YRS	3%	4%	5%	6%	7%	8%	9%	10%	11%	12%	15%
						INTEREST RATE						
0%	1	97	96	95	94	93	92	91	91	90	89	86
	2	94	92	90	89	87	85	84	82	80	79	74
	3	91	89	86	84	81	79	76	74	72	70	6
	4	89	85	82	79	76	73	70	67	64	62	55
	5	86	82	78	74	70	67	64	61	58	55	47
	6	84	79	74	70	66	62	58	55	52	49	41
	7	81	76	70	66	61	57	53	50	46	43	35
	8	79	73	67	62	57	53	49	45	42	38	30
	9	76	70	64	58	53	49	45	41	37	34	26
	10	74	67	61	55	50	45	41	37	33	30	22
	12	70	62	55	49	43	38	34	30	27	24	19
	15	64	55	47	41	35	30	26	22	19	17	12
15%	1	97	97	96	95	94	93	93	92	91	90	88
	2	95	93	92	90	89	87	86	84	83	81	77
	3	93	90	88	86	84	81	79	77	75	73	68
	4	97	87	84	81	79	76	74	.71	69	66	60
	5	88	84	81	77	74	71	68	65	62	60	53
	6	86	82	77	74	70	66	63	60	57	54	46
	7	84	79	74	70	66	62	58	55	52	49	41
	8	82	76	71	66	62	58	54	50	47	44	36
	9	79	74	68	63	58	54	50	46	43	40	31
	10	77	71	65	60	55	50	46	43	39	36	28
	12	74	66	60	54	49	44	40	36	32	29	21
	15	68	60	53	46	41	36	32	28	24	21	14
28%	1	98	97	96	96	95	94	94	93	92	92	90

TAX RATE	YRS	INTEREST RATE										
		3%	4%	5%	6%	7%	8%	9%	10%	11%	12%	15%
	2	96	94	93	92	90	89	88	86	85	84	80
	3	94	92	90	88	86	84	82	80	79	77	72
	4	92	89	87	84	82	79	77	75	73	70	64
	5	90	87	83	80	78	75	72	69	67	64	58
	6	88	84	80	77	74	71	67	65	62	59	52
	7	86	82	78	74	70	67	63	60	57	54	46
	8	84	79	75	71	67	63	59	56	53	50	41
	9	82	77	72	68	63	59	55	52	49	45	37
	10	81	75	70	65	60	56	52	48	45	42	33
	12	77	71	65	59	54	50	46	42	38	35	27
	15	72	65	58	52	47	42	37	33	30	27	19
31%	1	98	97	97	96	95	95	94	93	93	92	90
	2	96	95	93	92	91	89	88	87	86	84	81
	3	94	93	90	88	86	85	83	81	79	78	73
	4	92	90	87	85	82	80	78	76	73	71	65
	5	90	87	84	81	78	76	73	70	68	66	59
	6	88	85	81	78	75	72	69	66	63	60	53
	7	86	82	78	75	71	68	64	61	58	55	48
	8	85	80	76	72	68	64	60	57	54	51	43
	9	83	78	73	69	64	60	57	53	50	47	39
	10	81	76	71	66	61	57	53	50	46	43	35
	12	78	72	66	61	56	51	47	43	40	36	28
	15	73	66	59	53	48	43	39	35	31	28	20

3.3—Interest on Monthly Deposits in Your Savings Account

If you make regular monthly deposits into a daily-interest savings account, how much interest will you earn at the end of each year? Consult Table 3.3. (You'll find the mathematical basis of this table in C.6, page 324.)

 EXAMPLE **You want to deposit $1 per month at a daily-interest rate of 9%.**

If you have no taxes to pay on the interest, you will earn $73 after 10 years. At that time, the balance in your account will be $1 × 120 months = $120—plus the interest of $73, making $193.

As we explain in Section 3.1, pages 44 to 46, you can make adjustments to take account of the following changes:

- Amounts other than $1.
- Interpolating between values.
- Extending the number of years.
- Different compounding periods.
- Effects of inflation.

TABLE 3.3

Interest Earned, After Taxes, on Monthly Deposits of $1 in a Daily-Interest Savings Account (Dollars)

TAX RATE	YRS	INTEREST RATE										
		3%	4%	5%	6%	7%	8%	9%	10%	11%	12%	15%
0%	1	0.2	0.2	0.3	0.3	0.4	0.5	0.5	0.6	0.6	0.7	0.9
	2	0.7	0.9	1.2	1.4	1.7	2.0	2.2	2.5	2.7	3.0	4.8
	3	1.6	2.1	2.8	3.3	4.0	4.6	5.2	5.8	6.4	7.1	9.2
	4	2.9	3.9	5.1	6.1	7.3	8.4	9.6	11	12	13	17
	5	4.6	6.2	8.1	9.8	12	13	15	17	20	22	29
	6	6.7	9.1	12	15	16	20	23	26	30	33	44
	7	9.3	13	16	21	23	28	32	37	42	47	64
	8	12	17	22	28	31	38	43	50	57	64	88
	9	16	22	27	36	41	50	57	66	75	85	119
	10	20	27	33	45	52	64	73	85	97	111	157
	12	29	40	50	64	80	98	114	133	153	176	257
	15	47	66	87	111	137	167	199	236	276	322	494
15%	1	0.1	0.2	0.2	0.3	0.3	0.4	0.4	0.6	0.5	0.6	0.7
	2	0.6	0.8	1.0	1.2	1.4	1.6	1.9	2.1	2.3	2.5	3.2
	3	1.4	1.8	2.3	2.8	3.3	3.8	4.4	4.9	5.4	6.0	7.7
	4	2.7	3.3	4.2	5.2	6.1	7.0	8.0	9.0	10	11	14
	5	3.9	5.3	6.7	8.2	9.8	11	13	15	16	18	24
	6	5.7	7.8	10	12	14	17	19	22	24	27	36
	7	7.9	11	14	17	20	23	27	31	36	38	52
	8	10	14	18	22	27	31	36	41	47	52	71
	9	13	18	26	29	35	37	47	54	61	69	95
	10	17	23	30	36	44	52	60	69	79	89	123
	12	24	34	44	54	66	78	92	106	122	139	198
	15	39	55	72	90	110	138	157	184	214	248	367

TAX RATE	YRS	INTEREST RATE										
		3%	4%	5%	6%	7%	8%	9%	10%	11%	12%	15%
28%	1	0.1	0.2	0.2	0.2	0.2	0.2	0.3	0.4	0.5	0.5	0.6
	2	0.5	0.7	0.9	1.0	1.2	1.4	1.4	1.6	1.8	2.1	2.7
	3	1.2	1.6	2.0	2.4	2.8	3.2	3.7	4.1	4.6	5.0	6.5
	4	2.1	2.8	3.6	4.3	5.1	5.9	6.7	7.6	8.4	9.3	12
	5	3.3	4.5	5.7	6.9	8.2	9.5	11	12	14	15	20
	6	4.8	6.5	8.3	10	12	14	16	18	20	22	30
	7	6.6	9	11	14	17	19	22	25	28	32	42
	8	8.8	12	15	19	22	26	30	34	35	43	57
	9	12	15	19	24	29	34	39	44	50	56	76
	10	14	19	24	30	36	42	49	56	64	71	98
	12	20	28	36	45	54	64	74	85	97	110	154
	15	32	45	59	73	89	107	125	146	168	192	277
31%	1	0.1	0.2	0.2	0.2	0.3	0.3	0.4	0.4	0.4	0.5	0.6
	2	0.5	0.6	0.8	1.0	1.2	1.3	1.5	1.7	1.9	2.0	2.6
	3	1.1	1.5	1.9	2.3	2.7	3.1	3.5	3.9	4.4	4.8	6.2
	4	2.0	2.7	3.4	4.1	4.9	5.7	6.4	7.2	8.0	8.9	11
	5	3.2	4.3	5.4	6.6	7.8	9.0	10	12	13	14	19
	6	4.2	6.2	7.9	9.7	12	13	15	17	19	23	28
	7	6.3	8.9	11	13	16	19	21	24	27	30	40
	8	8.3	11	14	18	21	25	28	32	36	41	54
	9	11	15	19	23	27	32	37	42	47	53	72
	10	13	18	23	29	34	40	47	55	60	68	92
	12	20	27	34	43	51	61	70	81	92	104	145
	15	31	43	56	70	85	101	118	137	158	180	259

3.4—Monthly Deposits You Need to Reach a Goal

How much must you deposit each month into a daily-interest savings account to accumulate a certain balance at some time in the future? Consult Table 3.4. (You'll find the mathematical basis of this table in C.6, page 325.)

EXAMPLE

You want to accumulate $1,000 by the end of five years, and want to know how much to deposit.

At an interest rate of 9%, compounded daily, and while paying taxes at the marginal rate of 31%, Table 3.4 shows that you must deposit $14 per month.

As explained in Section 3.1, pages 44 to 46, you can make adjustments to take account of the following changes:
- Amounts other than $100.
- Interpolating between values.
- Extending the number of years.
- Different compounding periods.
- Effects of inflation.

TABLE 3.4

Monthly Deposits Needed to Accumulate a Future
Balance, After Taxes, of $1,000 with Daily
Interest Added (Dollars per Month)

TAX RATE	YRS	3%	4%	5%	6%	7%	8%	9%	10%	11%	12%	15%
						INTEREST	RATE					
0%	1	82	82	82	81	81	80	80	80	79	79	78
	2	40	40	40	39	39	39	38	38	37	37	36
	3	27	27	26	26	25	25	25	24	24	23	22
	4	20	20	19	19	18	18	17	17	16	16	15
	5	15	15	15	14	14	14	13	13	12	12	11
	6	13	13	12	12	11	11	10	10	9.7	9.5	8.6
	7	11	11	10	9.7	9.3	8.9	8.6	8.3	7.9	7.6	6.8
	8	9.2	8.8	8.5	8.1	7.8	7.5	7.1	6.8	6.5	6.2	5.4
	9	8.1	7.7	7.3	7.0	6.7	6.4	6.0	5.7	5.5	5.2	4.4
	10	7.2	6.8	6.4	6.1	5.8	5.6	5.2	4.9	4.6	4.3	3.6
	12	5.8	5.4	5.1	4.7	4.4	4.2	3.9	3.7	3.4	3.1	2.5
	15	4.4	4.0	3.7	3.4	3.2	2.9	2.6	2.4	2.2	2.0	1.5
15%	1	82	82	82	81	81	81	80	80	80	79	79
	2	41	40	40	40	40	39	39	38	38	38	37
	3	27	26	26	26	26	25	25	24	24	24	23
	4	20	19	19	19	19	18	18	18	17	17	16
	5	16	15	15	15	15	14	14	13	13	13	12
	6	13	13	12	12	12	11	11	11	10	10	9
	7	11	11	10	9.9	9.6	9.3	9.0	8.7	8.4	8.2	7.4
	8	9.4	9.1	8.8	8.4	8.1	7.8	7.6	7.3	7.0	6.7	6.0
	9	8.2	7.9	7.6	7.3	7.0	6.7	6.4	6.2	5.9	5.7	4.9
	10	7.3	7.0	6.7	6.4	6.1	5.8	5.6	5.3	5.0	4.8	4.1
	12	5.9	5.6	5.3	5.0	4.8	4.5	4.2	4.0	3.8	3.5	2.9
	15	4.6	4.3	4.0	3.7	3.4	3.2	30	2.7	2.5	2.3	1.8
28%	1	83	82	82	82	81	81	81	81	80	80	79
	2	41	41	40	40	40	39	39	39	39	38	37

TAX RATE	YRS	3%	4%	5%	6%	7%	8%	9%	10%	11%	12%	15%
						INTEREST	RATE					
	3	27	27	26	26	26	25	25	25	25	24	24
	4	20	20	19	19	19	19	18	18	18	17	17
	5	16	16	15	15	15	14	14	14	14	13	13
	6	13	13	12	12	12	12	11	11	11	11	10
	7	11	11	10	10	10	9.7	9.4	9.2	8.9	8.7	7.9
	8	9.5	9.3	9.0	8.7	8.5	8.2	7.9	7.7	7.5	7.2	6.5
	9	8.4	8.1	7.8	7.6	7.3	7.1	6.8	6.6	6.3	6.1	5.4
	10	7.5	7.2	6.9	6.7	6.4	6.2	5.9	5.7	5.4	5.2	4.6
	12	6.1	5.8	5.6	5.3	5.1	4.8	4.6	4.4	4.1	3.9	3.4
	15	4.7	4.4	4.2	3.9	3.7	3.5	3.3	3.1	2.9	2.7	2.2
31%	1	83	82	82	82	81	81	81	81	80	80	79
	2	41	41	40	40	40	39	39	39	39	38	38
	3	27	27	26	26	26	26	25	25	25	25	24
	4	20	20	19	19	19	19	18	18	18	18	17
	5	16	16	15	15	15	14	14	14	14	13	13
	6	13	13	13	12	12	12	11	11	11	11	10
	7	11	11	11	10	10	10	9.5	9.3	9.0	8.8	8.1
	8	9.6	9.3	9.1	8.8	8.5	8.3	8.0	7.8	7.6	7.3	6.8
	9	8.4	8.2	7.9	7.6	7.4	7.1	6.9	6.7	6.4	6.4	5.6
	10	7.5	7.2	7.0	6.7	6.5	6.2	6.0	5.8	5.5	5.3	4.7
	12	6.1	5.9	5.6	5.4	5.1	4.9	4.7	4.4	4.2	4.0	3.5
	15	4.7	4.5	4.2	4.0	3.8	3.6	3.4	3.2	3.0	2.8	2.3

3.5—Present Value of a Series of Monthly Payments

As defined in C.2, page 316, the "present value" of a series of future payments is the amount that you must invest today at a certain interest rate to equal the total future value of the series. A simple example is given in C.2 to illustrate the concept (page 316).

Use Table 3.5 to determine the results for different interest rates and times. (You'll find the mathematical basis of this table in C.6, page 325.)

 You win a lottery and are given the choice between five years of monthly $100 payments and a single payment today.

If you make a series of monthly deposits of $100 over a period of five years and they earn 10% interest, compounded daily, the present value of the series (based on the same 10% interest rate) is found from Table 3.5 to be $47 × 100 = $4,700. This means that it would be better to choose the single payment if it is larger than $4,700 even though the series of 60 $100 payments amounts to $6,000.

Table 3.5 covers cases in which monthly payments begin now and continue for some years. In more-complicated cases, the payments begin at some future time and then continue for a period of years. In these cases, you need to use Table 3.5 first to bring the series to a virtual present value for the time when the series starts; then, in a second step, you treat this sum as a single future amount for which the real present value is found in Table 3.2, page 50.

EXAMPLE Your lottery prize is five years of monthly
$100 payments, beginning three years from
now, and you want to know the real present
value.

The virtual present value of $4,700 is found as in the
previous example. This would then be used to enter Table
3.2, page 50, using 10% and three years, finding the value
74—which, when multiplied by 47, gives the final answer
$3,478 as the real present value in this case.

As explained in Section 3.1, pages 44 to 46, you can make
adjustments to take account of the following changes:
- Amounts other than $1.
- Interpolating between values.
- Different compounding periods.
- Effects of inflation.

TABLE 3.5

Present Value of a Series of Equal Monthly Payments of $1 (Dollars)

TAX RATE	YRS	DAILY INTEREST RATES										
		3%	4%	5%	6%	7%	8%	9%	10%	11%	12%	15%
0%	1	12	12	12	12	12	11	11	11	11	11	11
	2	23	23	23	23	23	22	22	22	22	22	21
	3	34	34	33	33	32	32	31	31	30	30	29
	4	45	44	44	43	42	41	40	39	38	37	36
	5	56	55	54	52	50	49	48	47	46	45	42
	6	66	64	62	60	58	56	55	54	52	51	47
	7	76	73	70	68	66	64	62	60	58	57	52
	8	85	82	79	76	73	70	68	65	63	61	56
	9	95	91	87	83	80	77	74	71	68	66	59
	10	104	99	95	90	86	82	79	75	72	70	62
	12	121	115	108	102	97	92	88	84	80	76	66
	15	145	135	126	118	111	104	98	93	88	83	71

TAX RATE	YRS	DAILY INTEREST RATES										
		3%	4%	5%	6%	7%	8%	9%	10%	11%	12%	15%
15%	1	12	12	12	12	12	10	10	10	11	11	11
	2	23	23	23	23	23	22	22	22	22	22	21
	3	35	34	34	33	33	32	32	32	31	31	30
	4	46	45	44	43	43	42	41	41	40	39	37
	5	56	55	54	53	52	51	50	49	48	47	44
	6	67	65	63	62	60	59	57	56	55	53	50
	7	77	75	72	70	68	67	65	63	61	60	55
	8	87	84	81	79	76	74	71	69	67	65	60
	9	96	93	90	86	83	80	78	75	73	70	64
	10	106	102	98	94	90	87	83	80	77	75	67
	12	124	118	113	107	103	98	94	90	86	82	73
	15	149	141	133	125	119	112	106	101	96	91	79
28%	1	12	12	12	12	12	12	12	12	11	11	11
	2	23	23	23	23	23	23	22	22	22	22	22
	3	35	34	34	34	33	33	33	32	32	32	31
	4	46	45	45	44	43	43	42	41	41	40	39
	5	57	56	55	54	53	52	51	50	49	48	46
	6	67	66	65	63	62	61	59	58	57	56	51
	7	78	76	74	72	71	69	67	66	64	63	58
	8	88	86	83	81	79	77	74	73	71	69	63
	9	98	95	92	89	86	84	81	79	77	74	68
	10	108	104	101	97	94	91	88	85	85	80	72
	12	127	122	117	112	108	103	99	96	92	89	79
	15	154	146	139	132	126	120	114	109	104	100	88

TAX RATE	YRS	DAILY INTEREST RATES										
		3%	4%	5%	6%	7%	8%	9%	10%	11%	12%	15%
31%	1	12	12	12	12	12	12	12	12	12	11	11
	2	23	23	23	23	24	23	22	22	22	22	22
	3	35	35	34	34	36	33	33	32	32	32	31
	4	46	45	45	44	44	43	42	42	42	41	39
	5	57	56	55	54	53	52	51	50	50	49	46
	6	68	66	65	64	62	61	60	59	59	56	53
	7	78	76	74	73	71	69	68	66	66	63	59
	8	88	86	84	81	79	77	75	73	73	70	64
	9	98	95	93	90	87	85	82	80	80	75	69
	10	108	105	101	98	95	92	89	86	86	81	74
	12	127	122	118	113	109	105	101	97	97	90	81
	15	155	147	140	134	127	122	116	111	111	110	90

3.6—Monthly Income That You Can Draw from a Capital Sum

If you have made a deposit in an interest bearing-account and want to draw a regular monthly income from it, Table 3.6 can be used to find how much you can withdraw over various numbers of years. (You'll find the mathematical basis of this table can in C.6, page 325.)

The table also shows how much you can withdraw without touching the original principal.

 EXAMPLE

You have $1,000 in a 9% daily-interest account and want to find how much per month you can withdraw over a period of five years.

If your tax rate is 28%, Table 3.6 shows that $20 per month can be withdrawn before you run out. If you wish to maintain the original principal of $1,000, you may withdraw only $5.40 per month.

As explained in Section 3.1, pages 44 to 46, you can make adjustments to take account of the following changes:
- Amounts other than $1,000
- Interpolating between values
- Different compounding periods
- Effects of inflation

TABLE 3.6

Monthly After-tax Income Drawn from $1,000 in a Daily-Interest Account (Dollars)

TAX RATE	YRS	INTEREST RATE										
		3%	4%	5%	6%	7%	8%	9%	10%	11%	12%	15%
0%	1	84	85	86	86	86	87	87	88	88	89	90
	2	42	42	42	43	44	45	45	46	46	47	48
	3	29	29	30	30	31	31	32	32	33	33	35
	4	22	22	22	23	23	24	24	25	26	27	28
	5	17	18	18	19	19	20	20	21	21	22	24
	6	15	16	16	16	17	17	17	18	18	20	22
	7	14	14	14	14	15	15	15	16	16	18	20
	8	12	12	12	13	13	14	14	15	16	16	18
	9	11	11	11	12	12	13	13	14	14	15	17
	10	9.6	10	10	11	11	12	12	13	13	14	16
	15	6.8	7.3	7.8	8.4	9.0	9.6	10	11	11	12	14
	*	2.5	3.3	4.2	5.0	5.8	6.7	7.5	8.4	9.2	10	13
15%	1	84	84	84	83	83	83	82	82	82	82	83
	2	42	42	43	43	43	43	43	44	44	45	46
	3	29	29	29	29	30	31	31	32	32	32	33
	4	21	22	22	23	23	24	24	25	25	25	26
	5	17	18	18	19	19	20	20	20	21	21	23
	6	15	15	16	16	16	17	17	18	18	19	20
	7	13	13	14	14	14	15	15	16	16	17	18
	8	11	12	12	13	13	13	14	14	15	15	17
	9	10	11	11	11	12	12	13	13	14	14	16
	10	9.4	10	10	11	11	11	12	12	13	13	15
	15	6.8	7.1	7.5	7.9	8.4	8.9	9.4	10	10	11	13
	*	1.9	2.6	3.6	4.5	5.1	5.6	6.2	7.1	7.8	8.6	11

* This amount of monthly withdrawal will enable the original principal to be maintained indefinitely.

TAX RATE	YRS	INTEREST RATE										
		3%	4%	5%	6%	7%	8%	9%	10%	11%	12%	15%
28%	1	84	84	84	84	84	84	84	84	87	87	88
	2	42	42	42	42	42	43	43	45	45	46	46
	3	28	28	29	29	29	29	30	30	31	31	32
	4	21	22	22	22	23	23	24	24	24	25	26
	5	17	18	18	18	19	19	20	20	20	21	22
	6	15	15	15	16	16	16	17	17	17	18	19
	7	13	13	13	14	14	14	15	15	16	16	17
	8	11	12	12	12	13	13	13	14	14	14	16
	9	10	10	11	11	11	12	12	13	13	13	15
	10	9.2	10	10	10	11	11	11	12	12	13	14
	15	6.5	6.8	7.2	7.6	8.0	8.3	8.8	9.1	9.6	10	11
	*	2.1	2.6	3.1	3.6	4.3	4.8	5.4	6.1	6.7	7.3	9.1
31%	1	84	84	85	85	86	86	86	86	87	87	88
	2	42	43	43	43	44	44	44	45	45	45	46
	3	28	29	29	30	30	30	30	31	31	31	32
	4	22	22	22	23	23	23	24	24	24	25	25
	5	17	18	18	18	19	19	19	20	20	20	21
	6	15	15	16	16	16	16	17	17	17	18	19
	7	13	13	13	14	14	14	15	15	15	16	17
	8	11	12	12	12	13	13	13	14	14	14	15
	9	10	10	11	11	11	12	12	12	13	13	14
	10	9.2	9.5	10	10	11	11	12	12	12	12	13
	15	6.5	6.7	7.1	7.5	7.8	8.2	8.6	9.0	9.4	9.8	11
	*	2.0	2.3	2.9	3.5	4.1	4.6	5.2	5.8	6.4	7.0	8.8

* This amount of monthly withdrawal will enable the original principal to be maintained indefinitely.

4

Health and Fitness by the Numbers

N umbers play a role in a great many health and physical-fitness topics, ranging from the probability that you will survive the hazards of daily life and live to a ripe old age, to the measures you can take—through diet and exercise, for example—to keep yourself healthy. This chapter covers a wide array of such numbers.

4.1—Life Expectancy

Life expectancy in the United States, for both men and women, has been increasing steadily during the past century. Figure 4.1 graphically shows the good news.

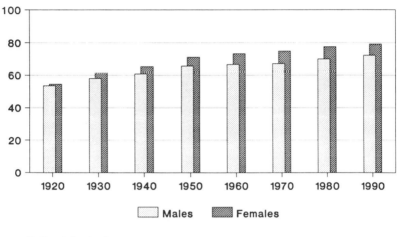

Source: National Center for
Health Statistics

Figure 4.1—U.S. Life Expectancy Increases since 1920

It is tempting to project this rising curve into the indefinite future—possibly at a slower and slower rate, but still trending upward. But the evidence is mounting that even if most killer diseases are conquered, the maximum human life expectancy will probably be about 85 years, according to a study by J. Olshansky and others that was reported in the journal *Science* on November 2, 1990. The human body degenerates naturally, and conquering major causes of death such as cancer and heart disease, which are primarily diseases of old age, will add only a few years to the average lifespan. In addition, life expectancy rates vary according to ethnic differences. For instance, blacks can expect to have a shorter life expectancy rate than whites—about seven fewer years for newborn males, and about five fewer years for newborn females, according to 1986 statistics from the U.S. National Center for Health Statistics.

◌▷ Numbers Are Not Forever

Statistics on life expectancies can be stated in many different ways: for males and females, for different races, for smokers and nonsmokers, for urban and rural residents, and so on. Commonly they are given, as in Table 4.1A, in terms of the expected number of years of life remaining if you are now at age 0, 10 years, 20 years, etc.

TABLE 4.1A

Remaining Life Expectancy for U.S. Males and Females (Estimated Figures for 1992)

Age in years	Male	Female
0	72.2 years	79.1 years
10	63.1	69.8
20	53.4	59.5
30	44.2	50.1
40	34.9	40.5
50	26.1	31.2
60	18.2	22.6
70	11.8	15.1
80	6.9	8.8
90	3.8	4.6

Based on 1988 data from National Center for Health Statistics, adjusted to 1992 estimates.

The table shows very clearly that at every age, life expectancy for men is consistently shorter than for women.

◌▷ "Quality Adjusted" Life Expectancy

Debilitating diseases frequently occur toward the end of life—which means that a measure of reasonably healthy lifespan will be less than the usual measure of life expectancy. Based on some estimates, the figures in Table 4.1A should be reduced by about five years for early years, scaling down fairly uniformly to about one year at age 90, to arrive at the average healthy lifespan before final illness begins.

☞ Percent That Will Die in a Year

Another way of expressing information similar to the life-expectancy figures in Table 4.1A is to show the percentage of a group of people of the same age who will die within the next year. Table 4.1B does just that.

TABLE 4.1B

Percentage of People at Different Ages Who Will Die Within the Next Year

AGE	MALES	FEMALES
1 to 4 years	.01%	.01%
5 to 14	.03%	.02%
15 to 24	.15%	.05%
25 to 34	.20%	.08%
35 to 44	.30%	.14%
45 to 54	.63%	.34%
55 to 64	1.60%	.90%
65 to 74	3.40%	2.00%
75 to 84	7.90%	5.00%
85 & over	17.60%	14.00%

Source: National Center for Health Statistics

Table 4.1B shows, for example, that if you are in the group of 65- to 74-year-old men, you may expect that 3.4% of the men in your age group will die within the next year.

☞ Influence of Heredity

The figures in Table 4.1A are averages for all males and females. You might want to make allowances for your own circumstances to get a better sense of your own life expectancy. It is impossible to do this precisely, but you can attempt an approximate estimate by considering how long your ancestors lived. If you find the figures for your parents, your grandparents, and so on, you can determine how many years each of them lived beyond the average for their time— or, how many years less than the average they lived. If you know that one of them died in war, or accidentally, you can leave him or her out of the averaging. To make this calculation, you can use a rough figure of 50 years for life expectancy at birth for the early years of this century and the latter years of the 19th century. In those years, life-expectancy differences between men and women were not significant. If you wish, include figures for your siblings and other relatives.

4.2—Life Insurance Costs for Smokers and Nonsmokers

The Surgeon General doesn't warn you about it, but it's true nonetheless: Smoking will cost you more than your health; it'll cost you plenty of money. Ask any life-insurance company.

☞ Insurance is Based on Hard Numbers

The figures below are those that insurance companies use to calculate the premiums they will charge. Some typical rates are given in Table 4.2 for male smokers and nonsmokers paying for the same $500,000 term-insurance coverage.

TABLE 4.2

Typical Insurance Premiums (Dollars)

AGE	RATES FOR NON-SMOKERS	RATES FOR SMOKERS	% INCREASE FOR SMOKERS
35	355	560	58%
40	460	720	57%
45	580	960	66%
50	775	1,295	67%
55	1,110	1,995	79%
60	1,940	3,235	66%
65	2,945	4,330	47%

Source: Orbita, from Average Quoted Rates

☞ The Numbers Tell You to Butt Out

Table 4.2 shows that for a typical life-insurance policy, smokers must pay somewhere between 50% and 80% more in premiums than nonsmokers. In this respect, there is no confusion about the cost of smoking—even if you are not convinced by some of the other statistical measures, such as reduction in life expectancy or "quality adjusted" life expectancy.

4.3—Blood Pressure Measurement

✏ Type-A Numbers

You go to your doctor's office and one of the nurse's first duties is taking your blood pressure. Detecting and controlling high blood pressure is especially important in preventing strokes, which occur as a result of a blockage or rupture of an artery in the brain.

When the left ventricle of the heart contracts and pumps blood through the main arteries of the body, there is a peak level of pressure exerted against the artery walls, which is called the "systolic" pressure. After the pumping contraction is complete, the pressure falls to a lower level, which is called the "diastolic" pressure.

The units used in these pressure measurements are millimeters of mercury, derived from the apparatus originally used to measure pressure in terms of the height of a column of mercury in a glass tube that would be supported by the pressure. The actual units used in this connection are unimportant, because the significant question is whether the measured levels are higher or lower than standard levels. Here we are concerned only with relative rather than absolute levels of pressure. The two pressure readings are expressed as a fraction: systolic over diastolic. For example, your blood pressure might be said to be "120 over 80," which would be written as 120/80.

 Average readings for young adults in good physical condition lie typically between 115 and 120 for the systolic pressure, and between 75 and 80 for the diastolic pressure.

A large difference in either of the readings from average values indicates some blood condition that should receive medical attention.

On the high side, readings higher than 140 for the systolic and 90 for the diastolic indicate a "high blood pressure" condition.

On the low side, readings below 100 for the systolic and below 60 for the diastolic indicate a "low blood pressure" condition.

4.4—Drinking and Driving

Recognizing the dangers of driving under the influence of alcohol, the U.S. Department of Transportation and many states distribute guidelines for determining your blood alcohol concentration. A useful general method is included here.

▷ Numbers on the Wagon

Driving ability is severely impaired at blood alcohol concentrations above .10 percent; even above .05 percent, driving ability is significantly affected because of reduced depth perception and increased reaction times. In many jurisdictions, you are considered legally impaired if your concentration exceeds .01 percent.

Table 4.4 can be used as an indicator of your ability to drive after drinking several drinks. The table is based on average rates of absorption of alcohol. One drink is assumed to be 12 ounces of beer, 5 ounces of table wine, 3 ounces of stronger wine, or 1.5 ounces of hard liquor (86 proof).

Table 4.4 - Drinks Before Driving

Body Weight				Number of Drinks in a 2-Hour Period						
100 Pounds	1	2	3	4	5	6	7	8	9	10
120	1	2	3	4	5	6	7	8	9	10
140	1	2	3	4	5	6	7	8	9	10
160	1	2	3	4	5	6	7	8	9	10
180	1	2	3	4	5	6	7	8	9	10
200	1	2	3	4	5	6	7	8	9	10
220	1	2	3	4	5	6	7	8	9	10
240	1	2	3	4	5	6	7	8	9	10

BE CAREFUL.	DRIVING IMPAIRED.	DO NOT DRIVE.
0% to .05%	.05% to .1%	over .1%

Source: US Dept of Transportation
National Highway Safety Administration

4.5—Cardiovascular Fitness

✏ Heart Rates by the Numbers

Experts suggest a formula for determining the the pulse rate you should maintain during aerobic activities such as running, biking or swimming. It says: Subtract your age from 220; find 72 percent of that figure for the low end of the range and 87 percent for the high end. Your pulse rate in beats per minute should stay between these two figures if you want to get the most cardiovascular benefit from your training. Figure 4.5 shows the values for different ages.

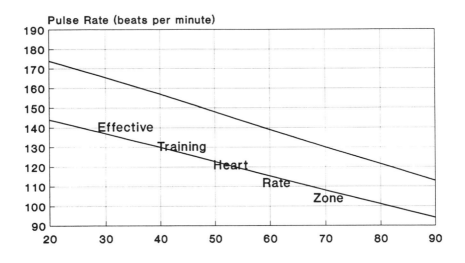

SOURCE: Ontario Association of Sport and Exercise Sciences

Figure 4.5—Heart Rate in Training

You are 20 years old, and you want to know what your ideal pulse rate is.

1: 220 − 20 = 200

2: 72 percent of 200 = 144

3: 87 percent of 200 = 174

Therefore, you should keep your pulse rate between 144 and 174.

4.6—Exercise Programs: How Often? How Hard? How Long?

☞ Active Numbers

Here is a "FITness Formula" developed by the Participaction Group, a Toronto-based organization that promotes increased physical activity and provides educational materials and films.

☞ "F" is for Frequency

How often you need to exercise is the first question. The answer is: at least three times a week. If more than 48 hours pass between sessions, the benefits begin to wear off. On the other hand, if you exercise more than five times a week, you risk overdoing it and feeling stressed and exhausted.

☞ "I" is for Intensity

How hard you should exercise is the second question. The answer is: hard enough to keep your heart rate in the target range for cardiovascular fitness. See 4.5, page 76. You don't need to adjust this range as your fitness level improves; it changes only with your age.

☞ "T" is for Time

How long you should exercise is the third question. The answer is: You must keep your heart rate in the desired range for at least 15 minutes continuously if you want to get the most out of that activity. If, however, your heart rate is above the desired range, it is important to back off: You're overdoing it.

☞ Exercising to Lose Weight

If you are dieting to lose weight and you want an exercise program to help you, there are many possible approaches. One that might work for you is a schedule of activities that can burn up about 2,000 calories a week. It involves doing one of the following:

Walking:	4 miles in an hour, five days a week.
Swimming:	30 minutes a day, six days a week.
Tennis:	1 hour a day, five days a week.
Jogging:	3 miles in 30 minutes, six days a week.

It is very difficult to lose weight by exercising alone. For example, a one-mile walk burns up only about 100 calories. The only practical program is a combination of diet and exercise. (See Section 4.8, page 83, for a discussion of diet and health.)

➯ Try an Informal Program

If you find it difficult to maintain a regular exercise program like that suggested above, you might have more success using a less structured program in which you increase your activity level by walking to work instead of taking the bus, using the stairs instead of the elevator, or not looking for the parking place nearest the door.

4.7—Healthy Body Weight

✏ Check the Numbers on Your Scales

At one time, health authorities put out lists that specified an ideal weight for a certain height. More recently, experts have begun to use a *range*.

The results, based on data from the U.S. Department of Agriculture, can be plotted in graphs such as Figure 4.7A for men and Figure 4.7B for women, which show bands of healthy weights for different heights. The results are valid for men and women between the ages of 20 and 65.

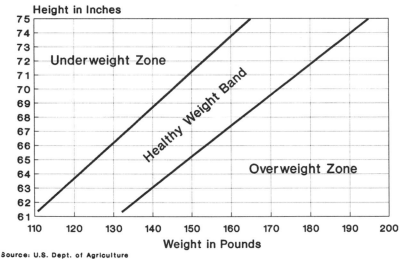

Source: U.S. Dept. of Agriculture

Figure 4.7A—Weight Chart: Men, 20–65

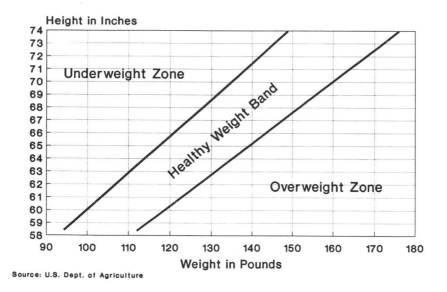

Figure 4.7B—Weight Chart: Women, 20–65

If you plot your weight and height and find that the result lies anywhere in the healthy-weight band, you may conclude that your weight is satisfactory. If it is to the left of the band, in the underweight zone, there is some cause for concern. If it is to the right of the band, in the overweight zone, you should begin to be concerned and should certainly not gain any more weight. If it is well into the overweight zone, there is an increased risk of health problems, including high blood pressure, heart trouble, etc. You should see a doctor or nutritionist and undertake some type of dieting and exercising program.

You should make this check using your height without shoes and your weight without clothes on. Probably the best time to take your weight is before breakfast in the morning.

Another indicator that you may find helpful is to recall what your weight was when you were in late teens or early twenties, when you were probably in reasonably good shape. This will indicate if you are naturally in the low or high end of the band for your height.

4.8—Diet and Health

🖘 Numbers that are Safe to Swallow

Few subjects have generated more controversy than what you should and should not eat. Books on diets, dieting, and health and fitness make up row after row of bookstore shelves. While many experts agree about quite a lot, there are many different theories on the subject. One thing is for sure—this is a topic where the picture is still being painted.

🖘 Dietary Balance

Your first concern in choosing a healthful diet should be with its balance. Many experts agree that you should get your calories from the basic types of food in the following proportions:

Protein... 12%

Fat ....20 to 30%

Complex carbohydrates.................53 to 63%

Sugars.....................................5 to 10%

(Complex carbohydrates are vegetables, fruits and grains; sugars are also carbohydrates, but in a simpler form. The other ingredients of food—fiber, water, vitamins, and minerals—are not included, since they don't contribute to its energy value.)

Balance is, apparently, hard to achieve. Americans typically get 40 percent of their calories from fat and 20 percent from sugars. If you can adjust your diet to achieve the recommended balance, you will be taking the most important step toward healthful eating.

Calories are units of energy value. One calorie is the energy needed to raise the temperature of one gram of water one degree Celcius. One pound of protein or one pound of carbohydrate has an energy value of about 1,500 calories, while one pound of fat has an energy value of about 3,300 calories—i.e., about 100 calories per tablespoon. Sugar has about 50 calories per tablespoon.

If you want to burn up calories by exercising, you should look in Section 4.6, page 78, where we suggested a program to use up 2,000 calories per week.

➡ Energy Needs

The daily energy requirements recommended by many experts to maintain health are shown in Table 4.8.

TABLE 4.8

Daily Energy Requirements

AGE (Years)	MALES		FEMALES	
	AVERAGE WEIGHT (lbs)	ENERGY NEEDS (Calories)	AVERAGE WEIGHT (lbs)	ENERGY NEEDS (Calories)
1	24	1,100	24	1,100
2–3	31	1,300	31	1,300
4–6	40	1,800	40	1,800
7–9	55	2,200	55	2,200
10–12	75	2,500	79	2,200
13–15	110	2,800	106	2,200
16–18	136	3,200	117	2,100
19–24	156	3,000	128	2,100
25–49	163	2,700	130	1,900
50–74	161	2,300	139	1,800
75+	152	2,000	141	1,500

Source: Adapted from National Research Council, Nutrition Board, and other sources.

➡ Excess Salt

Too much salt (sodium) in the diet causes extra water to be drawn into the blood vessels. This raises the pressure on artery walls, causing high blood pressure. You need about 1.5 teaspoons of sodium a day, but the average American takes in two to four times that much. In terms of milligrams, you need about 3,000 milligrams a day, so read labels carefully when you eat packaged food and put away the salt shaker you keep on the table.

➡ Excess Cholesterol

The problem of excess cholesterol in the blood is complicated because there are two forms—one harmful and the other, apparently, beneficial. The harmful kind is called LDL (for low-density lipoprotein), and the beneficial kind is called HDL (for high-density lipoprotein). Previously, you could have your blood tested for its cholesterol level and your doctor would tell you if your level was high. But the American Heart Association now recommends a more complicated test that separates the LDL from the HDL and tells you how you stand on each count. It can also tell you the ratio of the two types. However, the current thinking on the topic is not in absolute agreement, and the best you can do is cut back on animal fats in your diet—particularly red meat, butter, whole milk, cream, and fried foods. Stay alert for new numbers.

Vitamins and Minerals

It is difficult to provide definitive numbers that will clarify the complicated picture of your requirements for healthful levels of vitamins and minerals. This is an area that is full of hazards for the unwary, and where claims of manufacturers go far beyond reality. There are too many numbers out there. You can read the lists of ingredients in multiple vitamin and mineral supplements, all in micrograms and milligrams and obscure international units, which mean nothing in your everyday experience. There are U.S. RDAs (United States Recommended Dietary Allowances), set by the Food and Nutrition Board, which are shown on the labels of vitamin products. However, it is hard to find out how much of your RDA you are getting from your regular diet before you decide whether you need supplemental amounts.

Many registered dietitians say that if you are eating an adequate diet, with a balance of the four food groups (meat and fish, dairy products, fruits and vegetables, and bread and cereals), you are probably getting the vitamins and minerals you need. If you suspect you are lacking something, you will need to take the advice of a specialist you trust.

Do You Need Calcium?

There is one exception to the general statement that you are probably getting the minerals you need; it concerns the mineral calcium. Most experts recognize that it may be necessary to take special care with calcium. The body stores a large amount in the bones and teeth, which can be drawn on as a reserve for the small amount the body needs for its daily operations. But you can have difficulty maintaining the 500-to-800-milligrams-per-day level recommended for the average person and the 1,000-to-1,500-milligrams-per-day level for older women who may face a risk of osteoporosis. Probably the best source of extra calcium is skim milk, which has 300 milligrams per 8-ounce glass. Green vegetables are also high in this mineral. Again, if you suspect that you need a supplement, you should review your own diet and seek help from a specialist.

4.9—Sun Protection Factors (SPF)

✏️ Numbers from 1 to Tan

If you believe a suntan gives you that "healthy look," think again. Direct sunlight (ultraviolet radiation) is known to promote skin cancer.

With the risk of skin cancer on the rise, the U.S. Food and Drug Administration has set up a standard scale for sunscreens sold in the United States. This scale gives you some guidance to the effectiveness of these sunscreen lotions in protecting your skin from the harmful effects of the sun's rays. The SPF scale expresses the length of time you can stay in the sun with the lotion on, as compared to the length of time without it.

 EXAMPLE

If you can stay in the sun for 15 minutes without burning when you don't put on any lotion, then with a Number 8 lotion, you can stay for 15 × 8 = 120 minutes, or 2 hours. And with a Number 12 lotion, you can stay out for 3 hours.

There is some controversy over the effectiveness of these lotions, and certainly you must use them with care to ensure that you get complete coverage and that you maintain coverage at all times. Skin doctors are increasingly concerned with the long-term damage that results from exposure to the sun. Many say that you should use a lotion with at least SPF 15 whenever you are exposed to direct sunlight. Additionally, the sun is at its strongest between 10 A.M. and 3 P.M. ; most doctors advise people to minimize their exposure (even with sunscreens) during these hours.

4.10—20/20 and 14/14 Vision

✏ **Numbers with Corrective Lenses**

In assessing vision accuracy, the first test is to find out how well you can see objects at a distance. The first 20 in the expression "20/20 vision" refers to 20 feet, which is the distance between you and the chart your eye doctor wants you to read to see if you need corrective lenses—eyeglasses or contacts.

If, at a distance of 20 feet, you can read the line of printing on the chart that can be read at 20 feet by someone with normal eyesight, your doctor will tell you that you have 20/20 vision in that eye. But if, at a distance of 20 feet, you can read only a larger line of print that a normal eye can see from 30 feet, then your doctor will tell you that you have 20/30 vision in that eye. He or she will then try to correct your vision to bring it back to 20/20.

The second test reveals how well you see objects close-up. The first 14 in the expression "14/14 vision" refers to 14 inches, which is the distance between your eye and the card that your eye doctor uses, which approximates the distance at which you normally read a book. The same type of comparison made in the "20/20" test is made in this test. For example, if you can read only a line of large print that a normal eye can read at 36 inches, you would be told that your close-up vision is 14/36.

4.11—Spread of Diseases Such as AIDS

✏ Deadly Matters

Acquired Immune Deficiency Syndrome, or AIDS, was unknown, at least in the developed world, before 1981. According to estimates by the Centers for Disease Control in Atlanta, about 260,000 Americans will have died of AIDS by the end of 1992; and another 100,000 will have been infected with the AIDS virus, HIV.

The numbers indicate that the growth of this disease has been "exponential"—doubling at a rapid rate. There seems to be a "compound interest" factor at work, where the number of new cases grows as the number of existing cases grows.

The numbers need to be examined in relation to the exponential function to see if it can make them more understandable. The exponential function and its many uses are discussed in A.7, page 257.

Figure 4.11 is a plot of the cumulative total of AIDS cases in the United States, beginning in 1981 and with projections according to the estimated numbers that will be added in 1991 and 1992.

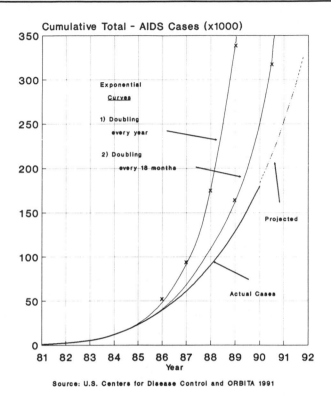

Figure 4.11—AIDS Cases in the United States

If you examine the number of cases in 1985 and 1986, you will see that the numbers were close to 21,000 and 39,000, respectively. So, during these years, the doubling time was just over one year.

Now if you look at the numbers for 1989 and 1991, you will see that the numbers are close to 130,000 and 250,000, respectively. These numbers show that, in these years, if the projections are accurate, the doubling time is just under two years.

The figure also shows an exponential curve beginning in 1985, and doubling every year, and a curve doubling every eighteen months.

⊃ Transmission Rates

If you look at Table A.7, page 260, you will see that if the doubling time is about two years, the Annual Percent Increase is about 35 percent—which means, in effect, that there is a 35 percent chance that every person with AIDS will pass on the infection to one other person during a one-year period.

That is a very rapid rate of transmission, and the only positive thing about it is that it may have come down from the corresponding transmission rate of about 60 percent in the 1985–86 period.

⊃ AIDS Numbers in the World

The World Health Organization is attempting to collect relevant AIDS figures, but, at present, they believe that the figures reported to them are no more than the tip of the iceberg. They estimate that up to 10 million people, worldwide, are infected with the HIV virus, which could mean that by 1993, up to 3 million cases of AIDS will develop from those who now carry the virus.

4.12—Biorhythms: Real and Imaginary

✏ Numbers on Bio-Cycles

Life is rhythmic at every level, and many numbers can be found to express the parameters of those rhythms.

- In the atoms and molecules that make up living cells, vibrations spread across an enormous range of frequencies. The electromagnetic spectrum includes the band of frequencies around 10^{15} hertz (or cycles per second) that we call visible light because it affects our eyes. It is an important rhythm of life. (See 8.11, page 197.)
- Our ears detect acoustic waves in a band of frequencies from about 20 to 20,000 cycles per second.
- Our heartbeats provide a central rhythmic pattern in our lives. The human heart beats in a band of frequencies from about 70 to 180 beats per minute.

✏ Slumber Numbers

A daytime, or diurnal, rhythm is set by the Earth's revolution around the sun. Our bodies respond to this 24-hour, or circadian, light/dark sequence with a sleeping/waking rhythm of their own. Many practical problems arise from this circadian rhythm, including those associated with shift work and jet lag. In some cultures, the siesta is favored because a 12-hour cycle seems to be more appropriate than a 24-hour cycle.

Studies show, however, that most humans have a natural daily cycle that is between 24 and 25 hours. Experiments have shown that humans have their own rhythms that have evolved independently of the sun's cycles. Societal time cues, though, keep us on a 24-hour cycle, and help us "reset" our body clocks each day.

✏ Lunar Numbers

The moon's motion sets rhythms that affect us in mysterious ways. Whether or not our moods and feelings are influenced by the moon is a controversial subject. Many attempts have been made to find a numerical backing for these suggestions—including, for example, counting the number of admissions to mental institutions, which seem to correlate with the phases of the moon.

The monthly rhythm that plays such a large part in human reproduction is no less mysterious. Does it represent some carry-over from an earlier time when the moon's influence was more direct? Or is it only a coincidence that these two rhythms have similar periodicities?

Solar Numbers

Seasonal rhythms were once of critical importance in human life. Now, their influence is only indirect, but it can still be felt, within us as well as in the outer world.

Many people suffer from a syndrome known as Seasonal Affective Depression (SAD), a feeling of gloom that occurs during the winter, when days shorten and nights get longer. A hormone called melatonin goes into the bloodstream through the pineal gland, an organ in the brain. Melatonin appears when it's dark; it goes away when it's light. It is involved in the regulation of internal rhythms and produces a sort of hibernation effect in humans. This "hibernation" often results in depression; therefore, an increase in light—even artificial light—can help those affected by SAD.

Are There Other Rhythms?

According to some believers, we are all subject to some other rhythms that many would say are imaginary rather than real. These other rhythms include 23-day "physical" cycles; 28-day "emotional" cycles; and 33-day "intellectual" cycles. Figure

To find your position on any given day with respect to these cycles, you have to find how many full cycles of each type you have passed through since you were born, and then the number of days you are "into" the current cycle. According to these theories, you are in an "up" period in the first half of the cycle and a "down" period in the second half. 4.12 shows these three cycles.

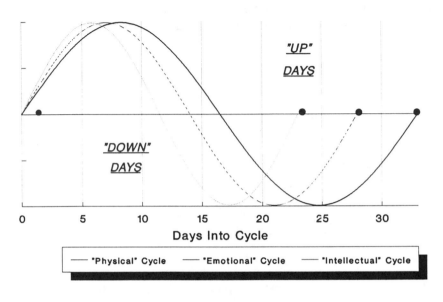

Figure 4.12—Biorhythms

The calendars in 9.5, beginning on page 219, will help you make a count of the number of days since you were born. Obviously, leap years are a complication in this process. When you find the total, you divide by the length of the cycle to get the number of cycles you have lived through, and the remainder tells you where you are in the current cycle.

Whether or not you believe that these cycles mean anything, it is an interesting exercise to calculate where you fit in the different cycles. If nothing else, it can be a lighthearted diversion—like consulting your horoscope.

4.13—Social Security Numbers

✎ Personal Numbers

You cannot think of aging and your health in today's world without reference to the basic number that identifies you as a unique individual: your Social Security number. On the one hand, you want it to be the key to accurate and comprehensive records of your personal situation, while, on the other hand, you may have an uneasy feeling that, in the wrong hands, it can reveal too much about you.

How did you get your number? When your application for a number is received, the Social Security Administration assigns the first three numbers according to a geographical coding. The second group of numbers is determined by the year in which your card is issued, according to a numerical sequence that is different for different states. The final four-number group is picked at random, which means that the complete number becomes a unique designator for you alone.

The facts are: Your number says something about where your application came from, and when it was received, but no other information is concealed in the coding.

5

Weather and the Environment

E veryone cares about the weather—whether our liveli-
hood depends on it, or merely our pleasure.

Our very survival depends on the health of our environment—
a subject that more and more of us have come to care deeply
about in recent years, as we've begun to see the potential
effects of rapid global changes.

As in so many other areas of life, numbers play a starring role.

5.1—Pressure and Weather Patterns

Air pressure is a vital determinant of weather. On weather maps, we see vast areas labeled "High" and "Low"—which indicate that within the irregular rings that mark their boundaries, the air pressure at the surface of the earth is either higher or lower than average, which is about 101.3 kilopascals. (Appendix B provides conversion factors for other units of pressure, but kilopascals are the units that have now been accepted for meteorological use.)

Owing to many causes—particularly atmospheric heating from below, the nature of the Earth's surface, elevation, and the season of the year—air pressure varies greatly from one air mass to another.

 The pressure at sea level in a low-pressure air mass in winter might be as low as 99 kilopascals, while the pressure at sea level in a high-pressure air mass in summer might be as high as 104 kilopascals.

The pattern and movement of these high- and low-pressure air masses, with their related weather systems and the warm and cold fronts between them, determine the weather conditions experienced at different locations.

Until the 1950s, weather forecasting was more art than science. But by mid-century, more and more numbers were being collected and rapidly plotted on weather maps, which began to show the changing patterns with greater accuracy. With the advent of powerful computers came the tantalizing possibility that if enough numbers could be collected and "crunched" quickly enough, computers might forecast weather more accurately than a skilled meteorologist. It was theorized that a giant computer fed with mountains of numbers that could be processed according to a set of empirical laws would yield reliable five-day, ten-day, or even longer range weather forecasts.

It was taken for granted that if meteorologists put enough effort into describing baseline conditions and identifying the physical laws of weather, they could predict the behavior of the atmospheric system with as much certainty as astronomers predict the motions of the planets using Newton's laws of motion. It was recognized that weather involved many more variables then planetary motion, so the problem would be more difficult, but it was nonetheless believed that the same basic methods should apply.

An implicit assumption was not recognized until much later. This was that a small error in describing the baseline conditions would produce some error in the final result, but this problem could be overcome by refining the baseline data until the final result was accurate enough to be useful. (This assumption was warranted in the case of planetary motions. Comets routinely return to the neighborhood of the Earth according to a schedule that can be worked out years in advance—and this occurs in spite of small errors in the initial conditions.) In short, it was assumed that nearly accurate inputs give nearly accurate outputs.

Like many others in the 1960s, meteorologist Edward Lorenz began to build computer models for analyzing weather. He knew that it was possible to predict the tides months in advance, and he hoped to build a weather model that would do the same.

He was plotting graphs of his results after entering the baseline conditions. Each set of consecutive measurements took some time to complete—so to save time, he decided to take a shortcut by starting a run partway through. To do this, he entered the values from the previous run as they had been at the point he was starting the second run. However, instead of duplicating the earlier results, the second run began to give different results. If the model were working properly, he would have expected it to repeat what had come out before. In a moment of illumination, Lorenz realized that when he put in the starting values for the second run, they were slightly different than before—because he entered only three places of decimals, while the earlier run was more accurate than that. This meant that the very small error in his input was being magnified, so that the final results were very different. In other words, his implicit assumption—that nearly accurate inputs give nearly accurate outputs—was wrong. Even tiny inaccuracies at any point in the process could radically alter the final result. In the jargon that came into use later, he recognized that the situation was "chaotic."

▱ Chaos is Normal

When he delved into the situation more carefully, Lorenz concluded that long-range weather forecasting is not merely difficult; it is impossible. Even if all the conditions at one instant could be exactly duplicated at a later instant, the subsequent pattern would quickly change in an unpredictable way. He found that chaos is normal.

Many years passed, and many millions of dollars were spent on supercomputers and satellite-linked data-collection systems that could give reasonably accurate two-day or three-day weather forecasts. But the hopes for better results quickly faded, and it was acknowledged that Lorenz was right. Forecasts for four or five days were useful indicators, but beyond six or seven days, the results were practically worthless.

The same picture of chaos began to emerge in many fields other than weather forecasting. In fact, a whole new science is developing that links many seemingly unrelated fields. Dynamic systems such as electronic oscillators, and biological and ecological systems such as animal populations, can show chaotic behavior. Chaos theory and the related subject of fractal geometry are emerging as topics of general interest. The 1990 bestseller *Chaos*, by James Gleick, is an excellent introduction. One reviewer asserted: "After reading *Chaos*, you will never look at the world in quite the same way again."

5.2—Temperature Scales

✏ Numbers with Diplomas

Temperatures measured in degrees Celsius (Centigrade) or degrees Fahrenheit may be converted using the well-known formulas:

$$C = (F - 32) \times \frac{5}{9}$$

or

$$F = \frac{9}{5}C + 32$$

To make the conversion mentally, a simple method to obtain Fahrenheit degrees if Celsius degrees are known is to double the Celsius value, subtract 10%, and add 32.

 EXAMPLE

To convert 24°C, you double it, 24 × 2 = 48, and subtract 10% of 48, or 5, to obtain 43. Adding 32 gives 75°F.

Some common equivalents are given in Table 5.2.

TABLE 5.2

Temperature Comparisons

F:	0	10	20	32	50	60	70	80	100	212
C:	−18	−12	−6.7	0	10	15.6	21.1	26.7	37.8	100

C:	−20	−10	0	10	20	30	40	100
F:	−4	14	32	50	68	86	104	212

5.3—Windchill Factors

⟣ Numb Numbers

When wind and a cold temperature come in tandem, the effect is a perception that it is much colder than the temperature alone would indicate. Wind lowers body temperature by evaporating perspiration or blowing away heat from the surface of your skin. To quantify this effect, a windchill factor has been devised—taking account of the cooling of an object exposed to the combined chilling effect of low temperature and wind. Table 5.3 gives the windchill temperatures in degrees Fahrenheit for a range of actual temperatures that are shown across the top of the table, and a range of wind speeds in miles per hour that are shown down the left side. (Note that winds in excess of 40 miles per hour produce little additional chilling effect.)

TABLE 5.3

Equivalent Windchill Temperatures

WIND SPEED MPH	DEGREES FAHRENHEIT													
0	30	25	20	15	10	5	0	−5	−10	−15	−20	−25	−30	−35
5	27	22	16	11	6	0	−5	−10	−15	−21	−26	−31	−36	−42
10	16	10	3	−3	−9	−15	−22	−27	−34	−40	−46	−52	−58	−64
15	9	2	−5	−11	−18	−25	−31	−38	−45	−51	−58	−65	−72	−78
20	4	−3	−10	−17	−24	−31	−39	−46	−53	−60	−67	−74	−81	−88
25	1	−7	−15	−22	−29	−36	−44	−51	−59	−66	−74	−81	−88	−96
30	−2	−10	−18	−25	−33	−41	−49	−56	−64	−71	−79	−86	−93	−101
35	−4	−12	−20	−27	−35	−43	−52	−58	−67	−74	−82	−89	−97	−105
40	−5	−13	−21	−29	−37	−45	−53	−60	−69	−76	−84	−92	−100	−107

Source: National Weather Service, U.S. Department of Commerce

5.4—Heat and Humidity Index

✏️ Summer Numbers

When high humidity combines with high temperature, it becomes very difficult for the body to lose enough heat to stay within its tolerance limits. Humidity refers to the amount of water vapor in the air. When humidity is high, the air feels *warmer* because the body's way of losing heat—sweating—is hindered as a result of the perspiration not evaporating as quickly in the moist air. When humidity is at lower levels, the air feels *cooler* than the actual temperature due to the efficiency of heat loss as a result of quick evaporation of perspiration.

A heat and humidity index has been drawn up to show how much hotter the average person feels when high humidity is combined with high temperature. Table 5.4 shows actual temperatures across the top and relative humidities down the left side. The table shows corresponding apparent temperatures.

TABLE 5.4

Heat and Humidity Index

Relative Humidity	Degrees Fahrenheit										
	70	75	80	85	90	95	100	105	110	115	120
0%	64	69	73	78	83	87	91	95	99	103	107
10%	65	70	75	80	85	90	95	100	105	111	116
20%	66	72	77	82	87	93	99	105	112	120	130
30%	67	73	78	84	90	96	104	113	123	135	148
40%	68	74	79	86	93	101	110	123	137	151	
50%	69	75	81	88	96	107	120	135	150		
60%	70	76	82	90	100	114	132	149			
70%	70	77	85	93	106	124	144				
80%	71	78	86	97	113	136					
90%	71	79	88	102	122						
100%	72	80	91	108							

Source: U.S. Department of Commerce, National Oceanic and Atmospheric Administration

The table shows, for example, that if the actual temperature is 90°F and the relative humidity is 60%, the apparent temperature—that is, what the temperature feels like—is 100°F. An apparent temperature above 105°F means that the body is in danger of suffering heat exhaustion.

5.5—Probability of Precipitation

✏ Numbers under Umbrellas

Daily weather forecasts normally include statements like, "The probability of precipitation today is 40%." This means that there is a 40% chance that there will be *some* precipitation *somewhere* in the forecast area *sometime* today. The statement does not mean that steady precipitation will fall over the whole area during 40% of the time, or that 40% of the area will have rain falling during the whole period.

In fact, this probability of precipitation is only an approximation; it cannot be precise, since it leaves open the questions of how much rain is meant by "some" rain, and how large an area is meant when it refers to "somewhere." Nevertheless, if the limitations are understood, the statement does provide a useful general-purpose measure of the likelihood that rain or snow will occur where you are during the period.

5.6—Holes in the Ozone Layer

✏ Tenuous Numbers

Our use of aerosol deodorants, hair sprays, air fresheners, and other products has created holes in the blanket of ozone that surrounds the Earth and shields us from ultraviolet radiation—the harmful component of sunlight.

Ozone (O_3) is a type of oxygen that you may have smelled if you were near an arcwelder or some other source of electric sparks. It is formed high in the atmosphere, where it absorbs the high-energy ultraviolet rays.

It is not easy to measure how much ozone the Earth's atmosphere contains, but recent measurements indicate that during the Antarctic summer, when the sun is shining almost continuously, the ozone level might be reduced by as much as 6%. The result is a "hole" that lets more ultraviolet light reach the surface of the Earth. The effect is more pronounced over the Antarctic because ice particles are present in the upper atmosphere that increase the rate of ozone depletion. The same processes apparently occur over the Arctic.

The cause of the worldwide reduction in ozone levels is the increased amount of atmospheric chlorine—resulting from the breakdown of chlorofluorocarbons (CFCs), which have been added to the atmosphere through their use in spray-can propellants, in refrigeration, in polystyrene products such as Styrofoam®, and in other products. Chlorine is a very effective agent for turning ozone to oxygen.

The ultraviolet light that reaches the Earth causes most skin cancers and also kills beneficial algae and bacteria that we depend on. World leaders have discussed agreements that would result in a gradual reduction in the use of CFCs. It appears that the problem can be solved, although it will mean using expensive substitutes for CFCs, some of which have not yet been found.

5.7—Air Pollution

⊟ Misty Numbers

In any one location, air quality may fluctuate wildly. One day, the air may pose no threat to the health of humans, animals or plants; the next, it may cause severe health problems. In recent years, governments have launched programs to measure pollution and air-quality levels—and to issue warnings or require action to reduce pollution levels when they reach a danger point.

⊟ EPA Surveys

In 1987, the U.S. Environmental Protection Agency (EPA) instituted nationwide surveys of air-pollution levels. The EPA found that more than 300 toxic chemicals were being released into the air, of which about 10 were of serious concern.

Two of the principal pollutants are sulphur dioxide (SO_2) and nitrogen dioxide (NO_2). About 20 million metric tons of each of these gases are being released every year, mainly from industrial plants that use coal or oil as a fuel, and from automobile exhaust. These two chemicals are the main causes of acid rain. (See 5.9, page 113.)

⊟ Air Quality Index

Other high-level pollutants are suspended particulates (that is, fine dust), carbon monoxide, and ozone. In some areas, the levels of these pollutants are measured continuously and an Air Quality Index (AQI) is issued. Concentrations are measured in parts per million (ppm), covering a range from 0 to more than 100 ppm's, divided into five categories.

Table 5.7 shows the categories and describes the effects of different levels of the pollutants.

TABLE 5.7

Air Quality Index

INDEX AND CATEGORY	CARBON MONOXIDE	NITROGEN	OZONE DIOXIDE	SULPHUR DIOXIDE	SUSPENDED PARTULATES
0–15ppm VERY GOOD	no effects	no effects	no effects	no effects	no effects
16–31ppm GOOD	no effects	slight odor	injures some plants	injures some plants	no effects
32–49ppm MODERATE	some blood effects	more odor	injures many plants	injures some plants	visi- bility reduced
50–99ppm POOR	cardio- vascular symptoms in smokers	some effects on asthmatic patients	reduces athletic perfor- mance	injures many plants	visi- bility severely reduced
100+ppm VERY POOR	cardio- vascular symptoms in non- smokers	more effects on asthmatic patients	affects patients with chronic pulmonary disease	affects patients with asthma & bron- chitis	affects patients with asthma & bron- chitis

Source: Based on EPA Data

When weather conditions cause pollution to accumulate in the air of some cities and the AQI rises into the Moderate range, industrial sources may be required to temporarily cut back their production until the air quality improves.

✏ Clean Air Act of 1990

Under provisions of the U.S. Clean Air Act of 1990, pollutant levels will be substantially reduced over the next few years. Some of the remedial steps to be taken require technical intervention to reduce emissions at their sources, while others involve energy conservation, use of alternate fuels, and the like. Regulating automobile exhaust is the principal way to reduce nitrogen dioxide pollution.

▷ Indoor Air Pollution

Energy conservation sometimes creates new hazards. Homes and offices are often so tightly sealed that today's air pollution indoors can be considerably worse than it is outside. Indoor air pollution can lead to ear, nose, and throat irritations, dizziness, nausea, fatigue, asthma, and other allergic reactions. The long-term effects can include kidney and liver diseases and various forms of cancer.

The harmful effects of tobacco smoke are becoming widely known. Other pollutants include asbestos fibers, mold, mildew, viruses, pet fibers, mites, formaldehyde, paints, solvents, sprays, air fresheners, disinfectants, and pesticide residues.

A particular cause for concern is radon gas—originating from radioactive materials in the soil, in rocks, in well water, and in some building materials. The maximum safe level of this gas in your home, according to the EPA, is 4 picocuries per liter of air. That is about three times the natural level of radon in the atmosphere.

5.8—Water Pollution

▷ Numbers in Your Drinking Glass

The main sources of water pollution are runoff from contaminated land, untreated or partially treated sewage, wastewater discharges from factories and industrial plants, improperly disposed toxic waste, and accidental chemical spills, particularly oil spills. (Acid rain as a source of water pollution is covered in 5.9, page 113.)

▷ Inland Water Pollution

In a major study completed in 1989, the EPA found that 595 inland water stretches, averaging six to ten miles long, were contaminated by harmful chemicals. The EPA named 126 chemicals causing concern. In addition, the agency identified more than 16,000 stretches of water that were contaminated by sewage and runoff water containing fertilizer and other wastes. The study concluded that about 10% of the nation's surface water is significantly polluted. The EPA went on to list 627 industrial plants, 240 local-government facilities, and 12 federal facilities (such as military bases) as the sources of toxic chemical pollution.

Following up on this study, U.S. Congress authorized the EPA to ensure that all these sources stop polluting by June 4, 1992, or face severe penalties.

▷ Ocean Pollution

Despite many efforts to reduce it, pollution of the ocean, particularly American coastal waters, has continued at unacceptable levels. The latest ban on ocean dumping enacted by Congress calls for complete cessation of this practice by 1992.

Before the Persian Gulf war in 1991, the number of major oil spills in the world's oceans was declining. The peak year, worldwide, was 1979, when about one million tons of oil were spilled. (Tons approximately equal one-seventh the number of spilled barrels; there are 42 gallons in a barrel.) Included in that total were 300,000 tons spilled from the *Atlantic Empress & Aegean Captain* near Trinidad

& Tobago and 36,000 tons spilled from the *Burmah Agate* in Galveston Bay, Texas. Over the past six or seven years, the annual worldwide total, on land and sea, has been less than one-fifth of the 1979 level. The *Exxon Valdez* spill in 1989 on the Alaskan coast was the last major spill. It polluted approximately 1,000 miles of coastline when 35,000 tons were spilled.

The situation in the Persian Gulf is without precedent. Massive amounts of highly toxic Kuwaiti crude oil released into the confined waters of the Gulf have presented the world with a clean-up problem that goes far beyond any earlier incidents. Even with expertise and resources assembled from many nations, it will take years to assess the full extent of the damage.

Underground Aquifers

Underground aquifers (a mass of sedimentary rock or gravel that produces water) provide about 60% of American drinking water, and water for irrigation through much of the central United States. For example: One huge aquifer underlies about two million acres of Texas, Kansas, and New Mexico. Water from underground aquifers is purified naturally as it is filtered through soil and permeable rock, but it can be polluted through spills, dumps, and other causes. Acid rain is one of those causes.

5.9—Acid Rain

"Acid rain" is the name commonly used to describe all forms of precipitation, including hail, snow, rain, and fog, that have become acidic because of various pollutants. (The general subject of air pollution is discussed in 5.7, page 108.)

All precipitation is slightly acidic because of dissolved carbon dioxide in the atmosphere, but this acidity can be greatly increased by oxides of nitrogen and sulphur in the air, which react with oxygen and moisture in the air to form dilute acids.

The pH Scale

The mathematical scale that is used to measure the acidity of any solution is called the "pH" scale. It is based on the concentration of hydrogen ions in the solution. For example: Vinegar, which is a dilute solution of acetic acid in water, has a hydrogen-ion concentration of 1/10,000 gram molecules per liter. Its pH value is said to be 4 (the number of zeros in the denominator of this fraction).

The neutral point, at which a solution is neither acidic nor alkaline, is 7, the midpoint on a scale running from 0 to 14. Any solution with a pH less than 7 is acidic; any with a pH greater than 7 is alkaline.

The major point to note about the pH scale is that a decrease of one pH value means a tenfold increase in acidity. For example: A solution with a pH value of 4 is 10 times as acidic as one with a pH value of 5.

The same pII scale is used to measure the acidity of soils. (See 8.10, page 196.)

pH Numbers Are Falling

Acid rain is a phenomenon of the modern industrial age. Snow that fell before the industrial revolution of the nineteenth century—preserved as ice in glaciers—has been found to have pH values generally above 5. But in the past century, the pH levels of precipitation have been declining—that is, the acidity has been rising—at an accelerating rate. Average rainstorm water falling over northeastern North America and over Western Europe is now about pH 4, indicating a tenfold increase in acidity since the industrial revolution. Extreme cases have shown levels around pH 2, which correspond to acidity levels 1,000 times pre-industrial levels.

5.10—Waste Disposal

⇨ The Garbage Crisis

Every day, the average American throws away three-and-a-half to four pounds of garbage. Something like 170 million tons of garbage must be picked up by municipal collection services. Most of that is household trash, but some industrial waste is included. That 170 million tons does not, however, include factory waste, construction and demolition waste, sewage sludge, medical waste, mine tailings, and scrapped cars and other machinery.

Over the past thirty years, the total trash to be collected has increased each year—and it will probably go on increasing over the next ten years by about 2% per year.

According to a study carried out in 1988 for the EPA by Franklin Associates Ltd., "Characteristics of Municipal Waste in the United States, 1960 to 2000," the composition of this garbage is roughly as shown in Table 5.10A.

TABLE 5.10A

Composition of U.S. Garbage (by Weight)

Food wastes	8%	Metals	9%
Yard wastes	18%	Rubber, leather,	
Newspapers	8%	textiles & wood	8%
Other paper products	33%	Plastics	6.5%
Glass	8%	Miscellaneous	1.6%

Source: U.S. Environmental Protection Agency

There are three ways to dispose of this mountain of garbage: Some wastes can be recycled; some can be burned to obtain energy and to reduce its volume; and the remainder must be taken to landfills.

⇨ Recycling

Recycling is the preferred method of disposal—more and more so as public awareness of the garbage glut grows and opposition to new landfills develops. Many communities are undertaking "curbside" programs to separate garbage into categories that will simplify recycling.

Newspapers, in particular, have been the focus of many recycling efforts. As of 1988, about 25% of all the newspapers trashed in the country were being recycled for re-use as newsprint, cardboard, insulation material, cat litter, and the like.

Some other materials are being recycled in useful quantities. They include other paper products (25% recycled); aluminum (25% recycled); ferrous metals (4% recycled); glass (9% recycled); and plastics (1% recycled).

✏ Incineration

Garbage-incineration programs can burn about 6% of the nation's garbage. The ashes weigh about 35% of the starting weight, which reduces the disposal problem. Some of the ashes can be recycled—for example, in road surfacing—but the problem of removing toxic materials from the ash has not been solved. Furthermore, the problem of air pollution caused by the smoke is a serious one that modern technology has reduced, but not eliminated.

✏ Landfill

About 80% of municipal garbage must still be disposed of by burying it under a layer of soil in a landfill. The garbage crisis has arisen because existing landfills are rapidly filling up and new sites are more difficult to find in the face of growing opposition from environmental groups and nearby residents.

✏ Plastics

Plastics present a special waste-disposal problem. Some ten million tons are discarded every year in the United States, of which only about one percent is recycled. The future promises an array of new efforts to recycle plastics, but the technical barriers are formidable.

✏️ **Hazardous Waste**

Some wastes are particularly troublesome because they contain hazardous materials. They include medical wastes, radioactive materials and hundreds of dangerous chemicals.

Medical wastes are usually incinerated, but they can still be hazardous. Recently, in the wake of medical wastes washing up on Atlantic shores, federal regulations have been tightened to reduce the risks to the public.

Radioactive materials, particularly those from nuclear-weapons and nuclear-power generators, represent a major disposal problem in the United States and many other countries. These wastes have been stored in temporary sites pending the development of secure, long-term sites—probably deep underground in geologically stable areas. This remains a problem for the future—for the long-term future—because some of the materials will remain radioactive for hundreds of thousands of years.

5.11—Electromagnetic Fields

⇒ Power Lines

The electromagnetic fields surrounding high-voltage power lines may harm the people living beneath them. Evidence suggests that the effects range from headaches to cancer. Some studies have suggested that minimum safe distances for prolonged exposure should be set at around 90 feet from 750-kilovolt lines, down to around 30 feet from 220-kilovolt lines; other studies have shown no measurable harm at closer proximity.

Lower-voltage lines—powering electrical vehicles and homes—are not thought to produce dangerous electromagnetic fields.

Because people get so close to them, the electromagnetic fields around the wiring in electric blankets and toasters are cause for concern. A federal panel reported in 1990 that leukemia in children may be caused by power lines, electric blankets, and toasters.

This is an area where reliable numbers are not available. The electrical activity in nerves, cells, the heart, and other muscles in the human body must be affected by external electromagnetic fields, but whether these extend to significant changes in human health is still an open question.

⇒ Microwave Ovens

One type of electromagnetic radiation that has a direct effect on tissue is present in microwave ovens. These operate at frequencies of either 890 to 915 megahertz or 2,400 to 2,500 megahertz. At these frequencies, molecules vibrate and generate heat. Evidence shows that exposure to low levels of microwave radiation can cause headaches, fatigue, and sleep disturbance.

Government safety standards have been set up that limit levels of radiation outside the oven to no more than five milliwatts per square centimeter at a distance of five centimeters from the door. Since the field strength diminishes in proportion to the square of the distance, this means that general levels in a room are extremely low if the safety standards are maintained.

5.12—Earthquakes

The most widely used measure of the magnitude of earthquakes is the numerical scale called the Richter Scale—named after Charles F. Richter, a California Institute of Technology seismologist who worked with Beno Gutenberg to perfect the scale. The instrument used to measure the ground motion and energy released in an earthquake is a seismograph. The scale of measurement is based on the seismograph reading that would be obtained by a seismograph located one hundred miles from the center of an earthquake (also known as the epicenter).

✎ Numbers with Bad Vibes

The Richter Scale is logarithmic, that is, based on a numerical system of exponents—in this case, ten. The scale measures the size of an earthquake in terms of the energy it expends. The scale starts at Magnitude 1, representing a nominal level of intensity. The next scale mark is 2, which represents a magnitude 10 times greater than 1. Scale mark 3 represents a magnitude 10 times greater again, or 100 times greater than 1, and so on. In general, an increase of one whole number, for example, from 6.0 to 7.0, represents an increase in the magnitude by a factor of 10. (This type of "logarithmic" scale is described further in A.5, page 251.) The scale is said to be "open-ended" because there is no upper limit on the magnitude that can be measured. (As a reference point, the famous 1906 San Francisco quake had an estimated Richter Scale reading of 8.3.)

Many thousands of earthquakes occur every year with magnitudes lower than 3.4 on the Richter Scale. These are of such low magnitude that they are not felt by people in the area where they occur.

Some sense of the relative magnitudes of earthquakes can be obtained by comparing their Richter Scale readings.

EXAMPLE The Mexico City quake of 1985 measured 8.1 on the Richter Scale, which means its magnitude was less than the 8.3 measure of the 1906 San Francisco quake by a factor that can be found from the ratio: $10^{8.3} \div 10^{8.1}$. The answer is 1.58, as shown by the methods explained in A.4, page 249, and A.5, page 251.

In terms of total energy released in different quakes, it is estimated that about 31 times more energy is released in a quake for every one-point jump on the Richter Scale.

The amount of damage an earthquake does depends on many factors. For example: The San Francisco quake of October 17, 1989, which was close in magnitude to the Armenian quake of December 1988 (6.9), caused fewer than 100 deaths, compared to more than 55,000 deaths caused by the Armenian quake. One theory for this difference in fatalities is that Armenian buildings did not meet with the same kinds of building safety codes as those in the United States. Also cited was the density of population in the stricken areas of Armenia, as compared to that in San Francisco.

5.13—Global Warming

➲ The Greenhouse Effect

The Earth's atmosphere traps the sun's heat in the same way a greenhouse does. Some sunlight reaching the Earth's surface is absorbed. The resulting heat is later emitted as infrared radiation, about 30% of which is absorbed by the atmosphere—mainly by carbon dioxide, but also by some of the polluting gases that are discussed in 5.7, page 108.

During the past century or so, the amount of carbon dioxide in the atmosphere has increased from .028% to .035%. Some scientists say that about two-thirds of the increase is due to burning fossil fuels (which generates CO_2) and the other third to deforestation (all plants absorb CO_2 as part of photosynthesis). The resulting increase in heat absorption occasioned by the increased CO_2 levels in the atmosphere has raised the average temperature of the Earth by about 0.9 degrees Fahrenheit.

Many scientists agree that the global temperature will increase over the next century, but estimates for the increase vary from 1.8 to 9.0 degrees Fahrenheit. Even that lower estimate would mean that the Earth's climate could change; the polar ice caps would melt, causing sea levels to rise, perhaps several feet; and changed amounts of rain would bring about many changes in the earth's vegetation.

5.14—Population Growth

▱ Exploding Numbers

The number of people in the world has been increasing so rapidly that overpopulation has become a threat to the planet's environment.

Table 5.14A shows the changes in U.S. population since 1900.

TABLE 5.14A

U.S. Population Growth

DECADE	POPULATION AT THE BEGINNING OF THE DECADE	AVERAGE ANNUAL GROWTH RATE
1900–1910	76.2 million	2.10%
1910–1920	92.3	1.50
1920–1930	106.0	1.62
1930–1940	123.2	0.72
1940–1950	132.2	1.45
1950–1960	151.3	1.85
1960–1970	179.3	1.34
1970–1980	203.3	1.14
1980–1990	226.5	1.02
1990–2000	249.6	0.7 (est.)

Source: U.S. Bureau of the Census

The table shows that U.S. population has more than tripled since 1900. The average annual growth rate has varied from about 2%, in the first decade of the century and during the "baby boom" years (1946–64), down to about 0.7% during the Depression years. The projected rate for the nineties is 0.7%, and this rate will probably fall to 0.5% in the following decade. In recent years, immigration has accounted for about 25% of the nation's annual population growth.

▭ Exponential Function

The exponential function discussed in A.7, page 257, can be used as a mathematical basis for analyzing these figures. Table A.7 gives doubling and tripling times for different growth rates.

Because the rate of growth of the U.S. population has fallen to a reasonable figure, it appears that this factor, by itself, is not a major source of environmental concern. A chance exists that a sustainable balance can be reached between population and resources.

▭ World Population Growth

At the global level, however, the situation is different. Table 5.14B shows rates of growth in the world.

TABLE 5.14B

World Population Growth

REGION	AVERAGE ANNUAL GROWTH RATE 1990–2000
World	1.7%
Developed World	0.5%
Developing World	2.0%
Sub-Saharan Africa	3.1%
Near East and North Africa	2.6%
Asia	1.7%
Latin America and the Caribbean	1.8%

Source: U.S. Bureau of the Census

A comparison of these figures to those in Table A.7, page 260, shows that the population of the developing world will double in about 35 years—and in certain parts of the developing world, particularly in Africa and the Near East, it will double in about 25 years.

The U.S. Bureau of the Census predicts that world population will go from its present level of 5.3 billion, to 6.3 billion by the year 2000, to 7.3 billion by 2010, and to 8.3 billion by 2020.

6

Gambling, Cards and Games—Odds and Probabilities

The numerical aspects of gambling, cards, and other games have been extensively studied over the years, with the principal objective of improving a player's odds of winning. Whatever your motivation, you will find that numbers can be an interesting component of these activities.

6.1—Gambling in General

Mathematical studies of gambling, particularly gambling with dice, trace a long history. One analysis of gambling was carried out by Italian astronomer and physicist Galileo more than 300 years ago.

👉 You Can't Beat the Numbers

Only a fool disregards the laws of probability when gambling. Betting on a long shot may occasionally be profitable, but over time, the number of heads will equal the number of tails in any coin-tossing situation—and similar statistical realities must be accepted in all games of chance.

Systems designed to beat the laws of probability, short of trickery, are impossible. In a few cases, systems may slightly improve your chances of winning, or, more realistically, reduce your chances of losing, but they can be successful only if they are designed to exploit the mathematical probabilities.

We will avoid the controversial subject of the extent to which skill can affect the outcome of gambling games. Clearly, rolling dice, spinning a roulette wheel, and pulling the handle of a one-armed bandit require no skill—but on the other hand, playing a complicated game like bridge requires a large element of skill. Somewhere between the two ends of the spectrum lie such games as poker and blackjack.

One general rule applies to all gambling games in which an individual plays a game against a casino or against an opponent with a much larger bank account: If two players sit down to an equitable game of chance (one in which each player has the same chance of winning) and the stakes are the same on each round, if the first player has 10 times the available capital of the second, then the odds are 10-to-1 that the second player will lose all his capital before the first.

👉 The House Has the Numbers

Furthermore, if the game is not equitable—that is, if the first player has even a slightly higher chance of winning each game, as well as more capital—the result is reinforced. This is the case in all casino games, where the house has a better-than-even chance of winning, as well as vastly more capital than any individual player.

You should bear this principle in mind any time you stake your limited capital against a richer opponent—even when the game is equitable. Your ultimate ruin is virtually certain, whether you play recklessly for high stakes or conservatively for small stakes. In all cases, the mathematical odds greatly favor the richer player.

6.2—Racetrack Betting

▭ Pari-mutuel Betting

The common form of at-the-track betting on thoroughbred races, harness races, and dog races is the pari-mutuel system—which, literally translated, means betting "among ourselves." In this system, all the money paid in at the betting windows is divided into "pools" corresponding to the different types of bets. After the track and the government each take a cut off the top, the money in each pool is paid out to those who hold winning tickets.

The usual types of bets are bets on a particular horse or dog to *win* a race; bets to *place* (that is, finish first or second); and bets to *show* (finish first, second or third). In addition, many other specialty bets can tempt the race enthusiast, including the exacta (first and second in correct order), the quinella (first and second in either order), and the daily double (winners in two consecutive races).

▭ Totalizator Numbers

Some time before a race is to be run, a set of approximate odds for each type of bet is flashed on a totalizator board—or "tote board." After a few minutes, the actual odds are calculated, based on the amount bet on each entrant in a race, and the total amount in the pool for that type of bet at that time. New sets of odds are recalculated and displayed at intervals of 90 seconds. Vending machines print and issue tickets as needed, and the totalizator sorts, adds and transmits the amounts held in each pool and the amount bet on each entrant in a race, from which new odds are calculated and displayed.

▭ Bookie's Numbers

When a bookmaker, or bookie, is involved (legally or illegally), a different form of betting is used. A bookie determines the odds and then takes in and pays out bets. In this case, the calculation of odds on horses in a race is a complicated and lengthy procedure involving many factors—including the track records of the horses, their ages, their pedigrees, their jockeys, the length of the race, the weather, the condition of the track, and the type and amount of the betting. In addition, a bookmaker quoting odds must make provision for his own expenses and any profit he hopes to make on the bets.

If a bookmaker judges a horse's chance of winning to be 2-to-1—that is, the bookmaker believes the horse has a 1-in-3 chance of winning—he might quote the odds as 3-to-2. Remembering that a winning ticket pays the original cost of the bet plus whatever profit the ticket earns at the final odds quoted, we observe that at 2-to-1 odds, a $2 ticket would pay the original $2 plus a $4 profit, making a total payout of $6 on a winner. At 3-to-2 odds, the payout would be $2 plus $3, or $5. In this case, the bookie would have $1 to cover his costs and profit.

▭ Payoffs or Prices

Knowing the odds quoted and recalling that a winning bettor receives the cost of the ticket as well as whatever profit the ticket earns at the final odds, we can determine the amounts horses will pay on the smallest bet accepted, usually $2.

If the final odds are 1-to-9, the so-called price or payoff (for a $2 bet) will be found by noting that 1-to-9 is the same as 2/9 to 2, and 2/9 = 0.22. Hence, the pari-mutuel price will be given as $0.22 plus the $2 for the ticket = $2.22, rounded down to $2.20. Similarly, odds of 9-to-1 equate to 18 to 2—giving a payoff of $18 + $2 = $20.

▭ Parlaying Bets

A common practice among bettors is to parlay bets—that is, to bet the winnings of one race on a later race. The following formula may be used to find the payoffs from successful parlays: Multiply the separate payoffs on the winning horses, divide by a factor, F (see below), and multiply by the dollars bet.

For a 2-horse parlay, $F = 4$

For a 3-horse parlay, $F = 8$

For a 4-horse parlay, $F = 16$

For a 5-horse parlay, $F = 64$

EXAMPLE

For a 2-horse parlay, if one horse paid $4.80 on a $1 bet and the other paid $10, then $4.80 × 10 = $48 and dividing by 4 = $12. Hence the payoff on the parlay is $12.

For a 3-horse parlay, if one horse paid $4, the second paid $7 and the third paid $20, then 4 × 7 × 20 = 560 and, dividing by 8 = $70. Hence, the payoff on the parlay is $70.

6.3—Lotteries

✏ Pull a Number Out of a Hat

If you consider only the numbers, you will never buy a lottery ticket. In almost every case, your chances of winning are so small that common sense should tell you to spend your money on something practical.

Yet millions of people buy lottery tickets every day—an indication that common sense is really not that common—or that people are paying for the pleasure they get from playing, rather than from any expectation of winning.

Interestingly, if the top prize in a lottery increases each time there is no winner, and eventually reaches, say, $20 million, far more people play than would play if the top prize were divided into 20 prizes of $1 million each. The odds of winning would, in fact, be better if the top prize were divided, but experience shows that people go for the bigger prize despite reduced odds. Intuitively, we can conclude that they choose to pay more for the greater pleasure of playing for a bigger prize.

When choosing between one lottery and another, many people choose the one that benefits a good cause or otherwise gives them pleasure. They don't make the choice through a comparison of their chances of winning.

If you do want to study the numbers, you can get a rough estimate of your chances based on counting the number of tickets that will be sold and dividing by the number of prizes. A simple example: If there is one prize and 1,000 tickets are sold, you have one chance in 1,000 of winning. But many real-life lotteries introduce other factors—such as split numbers, so that you win a smaller prize if you have the first three numbers correct, or the last three, either in the correct order or in any order. Sometimes it is not difficult to work out the chances of winning one of the smaller prizes by using permutations and combinations, which are reviewed in A.10, page 271. But it is often impossible to know how many tickets will be sold—and under these conditions, you simply cannot work out all the probabilities.

☞ Forget the Numbers and Enjoy

In the end, you may decide, as so many people do, to forget the numbers and just play for the fun of it. You can be sure of two things: Somebody is going to win, and you can't win (or lose) if you don't buy a ticket.

6.4—Blackjack

In this card game, aces can be counted as either 1 (a "low-ace" hand) or 11 (a "high-ace" hand). Face cards (jacks, queens, and kings) and 10s count as 10, and other cards count their face values. Each player—including the dealer—receives two cards, with one of the dealer's cards dealt face down. The goal is to beat the dealer, either by holding a hand that is higher than the dealer's (20 for you, 19 for the dealer) or by not "breaking" (going over 21)—at which point you automatically lose—and hoping the dealer doesn't break.

Adding up the value of your first two cards, you must determine whether you want to "stand pat" (not be dealt any more cards) or have the dealer "hit" you with another card. The goal is to accumulate cards totaling 21—or as close to 21 as possible, without exceeding it. If you hold an ace and a card valuing 10, you have 21, or blackjack. You win immediately. If you aren't dealt 21, and you wish to continue being hit, your goal is to have a higher score than the dealer—who always plays his or her hand last. Anytime your cards total 21, you win immediately. The dealer must continue being hit if his or hand is 16 or below. For a friendly, low-stakes game, here are some guidelines based on probabilities:

◉▷ **Rules for Friendly Games**

With a no-ace or low-ace hand...
- Stand pat (keep what you have) if your hand totals 17 or more.
- Draw (take a card) if your hand totals 11 or less.
- If your hand totals 12 to 16, stand pat if the dealer's up-card (the card whose value is showing) is low (2, 3, 4, 5, or 6); otherwise, draw.

With a high-ace hand...
- Stand pat if your hand totals 19 or more; otherwise draw.

When splitting pairs...

- If the game permits splitting pairs (two sets of two cards) to make two hands that are played separately, never split pairs other than aces, but always split aces. (The reasons are that for cards that count 10 and for 9's, their total of either 20 or 18 is probably already a winner, while for lower pairs, neither of the new hands starts off with favorable odds. Splitting the aces means that you have a good chance of getting two strong hands.)

▭▷ Winning Strategies are Possible

Unlike most gambling games, blackjack is a game in which skill can play a significant part. In fact, strategies can be developed that make it possible to win consistently in some casinos. However, such strategies are extremely complicated, requiring application of different rules for each different face-up card that is dealt. The skill involved is in learning to exploit the probabilities of each particular circumstance.

6.5—Rolling the Dice

Table 6.5A shows the 36 possible combinations that can be rolled in a single throw of a pair of dice, along with the corresponding totals.

TABLE 6.5A

Dice Scores

POSSIBLE COMBINATIONS						CORRESPONDING TOTALS					
(1,1)	(1,2)	(1,3)	(1,4)	(1,5)	(1,6)	2	3	4	5	6	7
(2,1)	(2,2)	(2,3)	(2,4)	(2,5)	(2,6)	3	4	5	6	7	8
(3,1)	(3,2)	(3,3)	(3,4)	(3,5)	(3,6)	4	5	6	7	8	9
(4,1)	(4,2)	(4,3)	(4,4)	(4,5)	(4,6)	5	6	7	8	9	10
(5,1)	(5,2)	(5,3)	(5,4)	(5,5)	(5,6)	6	7	8	9	10	11
(6,1)	(6,2)	(6,3)	(6,4)	(6,5)	(6,6)	7	8	9	10	11	12

Using the table of totals, it is easy to visualize the probabilities of rolling different scores. For example, there are six 7s in the table, meaning there are six ways in which a 7 can be rolled. Consequently, the probability of rolling a 7 is six out of 36, or one chance in six. (See A.10, page 270, for an outline of probability theory.)

Table 6.5B shows the frequencies of the totals.

TABLE 6.5B

Frequency of Dice Totals

TOTAL –	2	3	4	5	6	7	8	9	10	11	12
FREQUENCY –	1	2	3	4	5	6	5	4	3	2	1

Using this table, you can see that the probability of rolling a 7 can be found by dividing its frequency (6) by the total number of combinations (36).

✐ Summing Probabilities

According to the rule for summing probabilities (A.10, page 270), the probability of rolling one of a number of totals—say, a 2, a 3, or a

12—is the sum of the separate probabilities: $1/36 + 2/36 + 1/36 = 4/36$, or $1/9$.

✏ Multiplying Probabilities

According to the rule for multiplying probabilities (A.10, page 271), the probability of making a certain total and then of repeating the same total—say, two consecutive 7s—is the product of the separate probabilities: $6/36 \times 6/36 = 36/1,296$, or $1/36$.

✏ Making a Point

Variations in games of dice and methods of betting on them abound. In one common game, craps, you immediately win with a roll of 7 or 11; you immediately lose with a roll of 2, 3 or 12. If you roll any other total, that total becomes your "point." You then continue to roll until you either make your point and win your bet, or roll a 7 and lose your bet.

Table 6.5C shows the probability of rolling certain combinations.

TABLE 6.5C

Combinations of the Dice and the Probabilities of Rolling Them

SUM ON DICE	COMBI- NATIONS	DICE ROLLS	PROBABILITY
2	1	(1,1)	$1/36 = .028$
3	2	(1,2) (2,1)	$2/36 = .056$
4	3	(1,3) (3,1) (2,2)	$3/36 = .083$
5	4	(1,4) (4,1) (2,3) (3,2)	$4/36 = .111$
6	5	(1,5) (5,1) (2,4) (4,2) (3,3)	$5/36 = .139$
7	6	(1,6) (6,1) (2,5) (5,2) (3,4) (4,3)	$6/36 = .167$
8	5	(2,6) (6,2) (3,5) (5,3) (4,4)	$5/36 = .139$
9	4	(3,6) (6,3) (4,5) (5,4)	$4/36 = .111$
10	3	(4,6) (6,4) (5,5)	$3/36 = .083$
11	2	(5,6) (6,5)	$2/36 = .056$
12	1	(6,6)	$1/36 = .028$

6.6—Draw Poker

The game of draw poker may be played either with 52 cards, with a Joker added as a wild card (and 53rd card), or with 52 cards and the deuces wild. The object of the game is to wind up with a higher value of (five) cards than your opponents (or, if you're betting, to bluff your way into making your opponents *think* you have a better hand than they do), forcing them to fold before you reveal your hand.

Some key poker terms include straight flush: five cards in suit and sequence; royal flush: ace-king-queen-jack-ten of same suit; flush: five cards of same suit not in any sequence; full house: three cards of one rank and two of another; straight: five cards in sequence but in no particular suit; and wild card: card that can represent any desired card.

On the first deal of five cards, the chances of receiving different hands are shown in Table 6.6A.

TABLE 6.6A

Poker Hand Odds

HAND	52 CARDS	JOKER WILD	DEUCES WILD
1 chance in:			
5 OF A KIND	—	220,745	3,867
STRAIGHT FLUSH	64,974	14,067	5,370
4 OF A KIND	4,165	920	638
FULL HOUSE	694	438	205
FLUSH	509	368	197
STRAIGHT	255	140	39
3 OF A KIND	47	21	7.3
TWO PAIRS	27	23	21
ONE PAIR	2.4	2.3	2.1

These results can be extended to calculate the chances of being dealt a hand that has, for example a full house or better; a flush or better; a straight or better; and so on down the scale.

These chances are summarized in the Table 6.6C.

TABLE 6.6B

Number of Deals Required Before Certain Poker Hands Turn Up Once (Theoretically)

HAND	52 CARDS	JOKER WILD	DEUCES WILD
Full House or better	580	290	53
Flush or better	270	162	42
Straight or better	132	75	20
3-of-a-kind or better	35	16	6
Two pairs or better	13	10	5
Pair of Aces or better	9	7	4
Pair of Kings or better	7	6	4
Pair of Queens or better	6	5	3
Pair of Jacks or better	5	4	3
One pair or better	2	2	2

The minimum strength required to open the betting, and the minimum strength required to stay in after the betting starts, based on probabilities, are as shown in Table 6.6B.

TABLE 6.6C

Poker Strength to Open and to Stay

PLAYERS STILL TO BET	STRENGTH REQUIRED TO OPEN	PLAYERS STILL IN	STRENGTH REQUIRED TO STAY
1	ACE-KING HIGH	1	KINGS
2	PAIR OF 8's	2	2 LOW PAIRS
3	PAIR OF 10's	3	QUEENS UP
4	PAIR OF QUEENS	4	KINGS UP
5 OR 6	PAIR OF KINGS	5	ACES UP
7	PAIR OF ACES	6	3-OF-A-KIND

When deciding whether or not to draw cards to complete various poker hands, it is helpful to know the chances involved. Here are some common situations and the action you should take:

⮑ The Numbers Tell You What to Do

- When drawing three cards to a pair, the chances of making any improvement are only 1-in-2.5. The separate chances are 1-in-5 of making two pairs; 1-in-8 of making 3-of-a-kind; 1-in-100 of making a full house; and 1-in-360 of making 4-of-a-kind.

- When drawing two cards to 3-of-a-kind, the chances of making any improvement are only 1-in-9. The separate chances are 1-in-15 of making a full house and 1-in-22 of making 4-of-a-kind.

- When drawing one card to 3-of-a-kind, the chances of making any improvement are only 1-in-11. The separate chances are 1-in-15 of making a full house and 1-in-46 of making 4-of-a-kind.

- From the previous two points you can see that when holding 3-of-a-kind, the chances of improving the hand are slightly better if you draw two cards than if you draw only one.

- When playing with deuces wild, with five or six players in the game, the average winning hand will be three aces or higher, including one wild card. A combination with one or more wild cards is better than one without because it reduces the number of wild cards available to opponents. A good rule is: Never stay in without at least one wild card unless you hold 3-of-a-kind or better.

6.7—Roulette

Virtually every casino in the world features a spinning roulette wheel with its red and black spokes and its little bouncing ball.

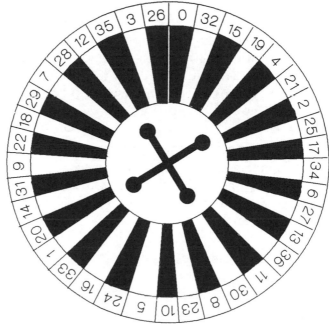

ORBITA 1991

Figure 6.7—The Roulette Wheel

In the United States, the wheel is divided into sections numbered 1 to 36, plus one marked 0 and another marked 00, making a total of 38. In Europe, the 00 section is left out and the total is 37.

To play, you place your bet on a board that is marked to indicate your choice.

✏ You Choose the Odds

Your choices include the following:
- a number;
- a color (black or red);
- even or odd;
- in the lower half (1 to 18) or the upper half (19 to 36) of the numbers;
- in the bottom dozen (1 to 12), the middle dozen (13 to 24) or the top dozen (25 to 36);
- either one of a pair of numbers, or 3 or 4 or 6 numbers.

When a ball settles on a zero, the house collects the stakes on all the other numbers. Since there are two zero sections on U.S. wheels, the odds favoring the house are 2 out of 38, or 5.26%. In Europe, with a single zero, the odds favoring the house are 1 out of 37, or 2.70%.

✏ The House Controls the Payoff

For each of the other selections, the payoff is set by the house to give itself a small edge—and, therefore, despite many efforts to devise a winning system, it is mathematically impossible to beat these odds over the long run. The only successful system is to quit when (and if) you are ahead, and never go back!

6.8—Bridge

The object of bridge is for a partnership to discover as much as they can about their combined cards and to arrive at a "contract" (winning a certain number of points, or tricks). Having "declared" their contract, an opposing partnership tries to stop them. The detailed rules and mathematical aspects of bridge are dealt with in many books. Only a few of the most interesting numbers can be covered here.

✏ Suits Me Fine

When you are dealt a hand, the probabilities of having certain distributions of the four suits in your hand can be worked out using the methods of probability theory outlined in A.10, page 270. The probabilities are listed in descending order in Table 6.8A. You may be surprised by some of the results.

TABLE 6.8A

The Ten Most Common Bridge Hands

No. of Cards per Suit	Probability	No. of Cards per Suit	Probability
(1) 4-4-3-2	0.215	(6) 6-3-2-2	0.056
(2) 5-3-3-2	0.155	(7) 6-4-2-1	0.047
(3) 5-4-3-1	0.129	(8) 6-3-3-1	0.034
(4) 5-4-2-2	0.106	(9) 5-5-2-1	0.032
(5) 4-3-3-3	0.105	(10) 4-4-4-1	0.030
Total (1) to (5)	0.710	Total (5) to (10)	0.200

You can see that the probability of getting one of the five most likely distributions is 0.71, (that is, about seven times in 10)--and, remarkably, that there is a 91% probability you will be dealt one of the top 10 distributions.

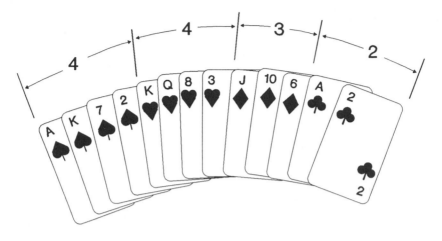

ORBITA 1991

Figure 6.8—The Most Common Hand—The 4-4-3-2 Distribution

Since almost one hand in four has the 4-4-3-2 distribution, one of the four hands dealt for every game, on the average, will have this distribution.

You will see this type of hand yourself, on the average, every fourth or fifth hand you play.

Another interesting set of distributions and their probabilities is shown in Table 6.8B.

TABLE 6.8B

Probabilities of Some Other Distributions

Distribution	Probability
one, two or three 4-card suits	0.35
one or two 5-card suits	0.44
one or two 6-card suits	0.17
one 7-card suit	0.035
one 8-card suit	0.005

From these results, you can see that:

- one out of three times, you can expect to be dealt a hand with one, two, or (rarely) three four-card suits;
- about half the time, you can expect to get one or (rarely) two five-card suits;
- the chances of getting a six-card-or-longer suit are less than 1-in-5;
- the chance that one of the suits will be void is only 1-in-20;
- the chance there will be a singleton in one of the suits is 1-in-3;
- there is a 50-50 chance of having a two-card suit.

No-trump Numbers

A "balanced" hand is often defined as a hand with no voids (no cards of a given suit) or singletons (one card of a given suit) and no more than one doubleton (two cards of a given suit). There are only three balanced hands, (4-4-3-2, 5-3-3-2, and 4-3-3-3), and it is interesting to see that all three are among the five most common hands. Collectively, they have a probability of 0.475. This means you have almost a 50-50 chance of having a balanced hand, and, since these are hands that play best in no-trump, you should often consider a no-trump bid.

Dummy Numbers

Table 6.8C gives the probabilities of finding various numbers of trumps (a particular card that ranks higher than any other plain card) in the dummy (the hand laid down by the declarer's partner and played by the declarer).

TABLE 6.8C

Trumps in the Dummy

TRUMPS IN BIDDER'S HAND	PROBABILITY OF FINDING TRUMPS IN THE DUMMY				
	ONE TRUMP	TWO TRUMPS	THREE TRUMPS	FOUR TRUMPS	MORE THAN 4 TRUMPS
4	(.12)	(.24)	.27	.21	.15
5	(.16)	.27	.27	.17	.09
6	.20	.31	.26	.13	.04
7	.26	.33	.22	.08	.02
8	.33	.33	.17	.04	.00

Another situation of interest is when you are playing the hand and, jointly with the dummy, you hold seven trumps. Then:

- if you lead trumps three times, you have a 36% chance of clearing the trumps from both of your opponents' hands;
- if you lead trumps four times, this chance goes up to 84%;
- if one honor card (the ace, king, queen, jack, or ten of a trump suit) is out, it will fall with two leads 18% of the time, and with three leads 54% of the time;
- if two honors are out, one will fall with one lead 8% of the time, and with three leads, both will fall 50% of the time.

✏ Random Shuffling

Sometimes people wonder: How many times must a deck of cards be shuffled to ensure that the distribution of the cards in the deck is truly random? Computer analysis indicates that the answer is about seven times, assuming that the average shuffling is not very thorough. Anything less than seven shufflings will leave some non-random patterns in the deck; anything more will not add very much to the randomness of the distribution.

6.9—Billiards, Snooker, and Pool

⇨ Straight Billiards

Straight billiards is played on a table with no pockets, using three balls: a white cue ball, another white ball marked with a spot, and a red ball. Points are scored for "cannons," when the cue ball touches both of the other balls.

⇨ Pocket Billiards

In one version of billiards, played on a table with pockets, three balls are used, as for straight billiards, and points are scored for putting the white ball and the red ball into a pocket. In another version, 15 colored balls, numbered from 1 to 15, and a cue ball are used. There are various ways of scoring.

⇨ Pool

Pool is played on a table with pockets, using fifteen colored balls, numbered 1 to 15, and a cue ball. A player tries to pocket a colored ball, making the points marked on the ball. The game ends when all the colored balls have been pocketed.

⇨ Snooker

Snooker is played on a table with pockets, using fifteen red balls, six colored balls, and a cue ball. A player tries to pocket a red ball, then a colored ball. At this stage in the game, if a colored ball is pocketed, it is set up again. When all the red balls are cleared, the colored balls are sunk in order: yellow (2), green (3), brown (4), blue (5), pink (6), and black (7).

7

Sports—Scoring and Statistics

U nderstanding sports is greatly enhanced when you understand the vast array of statistics you can find in the sports pages of your daily newspaper. This chapter discusses those numbers—except for the largest numbers of all, the salaries of professional athletes. No one understands those.

7.1—Golf

⇨ If Numbers Are Low, You Could Be a Pro

In addition to the simple arithmetic involved in scoring, several other numerical concepts are part of golf, including systems for handicapping, which allow players of differing abilities to compete fairly against one another; systems for adjusting course ratings so that scores are comparable between courses; and different methods of scoring matches and tournaments.

⇨ Finding Your Handicap

The basic method of calculating a player's handicap is to establish an average score, usually by keeping a running record of the best ten scores in the last twenty rounds of play. Subtracting par for the course from this average score establishes the player's handicap.

 EXAMPLE On a par 71 course, a player with an average score of 89 would be assigned a handicap of 18.

Sometimes the course rating rather than par is used in calculating handicaps. (See below.)

⇨ Equitable Stroke Control Procedure

An "equitable stroke control procedure" is often used during scoring to reduce the effects of unusually high scores on some holes.

The following limits are set on the hole scores used in calculating scores involving handicaps:

1: 1-over-par is recorded for "scratch" players, that is, players with 0 handicap.

2: 2-over-par is recorded on as many holes as the handicap exceeds scratch up to an 18 handicap, and 1-over-par on the balance of the holes. (For example, if you have a 15 handicap, you would be allowed to record a maximum score of 2-over-par on fifteen holes and 1-over-par on the other three.)

3: 3-over-par is recorded on as many holes as the handicap is increased over 18, and 2-over-par on the balance of the holes. (For example, if you have a 21 handicap, you would be allowed to record a maximum of 3-over-par on three holes and 2-over-par on the remaining fifteen.)

➯ The Slope System

In order to take into account differences between courses, a comparative rating system can be used. This may be a simple system based primarily on the length of holes, or a much more complicated system such as the "slope" system adopted in 1987 in the United States, which takes into account course obstacles as well as the length of holes. In either case, in tournaments the course rating is used with individual players' handicaps to arrive at final adjusted tournament scores.

When the U.S. Golf Association worked out and accepted the slope system, it did so after years of analysis. A team of raters, representing a range of golfing abilities, established a scale of obstacle-playing difficulties with which they were able to derive a national average of golf courses in the United States, having a slope of 113.

EXAMPLE If you have a handicap of, say, 4 on your home course (with a slope greater than 113), this might be adjusted to 2.8 on the standard course with a slope of 113. Another player might have a handicap of 18 on his or her home course (with a slope less than 113), which might translate to a handicap-slope index of 19.7. If you were playing the other player on a third course with a slope rating of 126, you would consult a table, which might indicate that your re-adjusted handicap is 3, while the other player's is 22. Therefore, in scoring that round you would give your opponent 22 −3 = 19 shots.

✏ Types of Matches

Matches can be played in many different ways. A match between two players is called a "single." Three players can play a "three-ball" match in which each player matches his or her score against each of the other two. In a "four-ball" match, two players are matched against another two; on each hole, the better ball of one pair is matched against the better ball of the other pair. A "Scotch foursome" is the same except that each pair plays only one ball, with the players taking turns.

✏ Tournament Scoring

Scoring in tournaments may be done using either "match" play or "stroke" play (sometimes called "medal" play). In either case, the play may be "at scratch" (that is, ignoring handicaps) or with handicaps taken into account.

In match play at scratch, each hole is a unit that is won by the player with the fewest strokes on the hole. The winner of the match is the player who wins the most holes in the match, regardless of his or her gross score for the round. Tied holes are said to be "halved," and they are not counted in the scoring. The match can finish early if one player is ahead by more holes than the number of holes left to play. A tie at the end of the round is broken by sudden-death play until one player wins a hole.

In handicap match play, the player's handicaps are considered at the beginning of the match and individual hole handicaps are calculated.

 If you have a handicap of 9 and your opponent has a handicap of 5, the difference is 4 and therefore you receive a one-stroke advantage on the four most difficult (regardless of par).

In stroke play, the winner is the player with the fewest total strokes for the 18 or 36 or 72 holes played. As in match play, the scoring can be either at scratch or with handicaps applied. Usually, handicaps are counted in this type of play. If it is necessary to play extra holes in a sudden-death playoff to break a tie, individual-hole handicaps are used as in handicap match play.

7.2—Tennis

✐ Love Those Numbers

The three aspects of tennis scoring concern games, sets, and matches. In the description to follow, singles play is described—that is, one player playing against a single opponent. But scoring is the same for doubles play, in which there are two players on each side.

✐ Scoring the Game

A point is scored when a player fails to hit the ball inside the far half of the court. The first point scored is termed "15." The second point is scored as "30" ; the third, as "40." These are not numerical point values in any sense, merely conventional names for the points. The term for a no-points score is "love." This term is an anglicized version of the French expression *l'oeuf*, meaning "the egg"— because a zero is egg-shaped. If the server wins the first point, the score becomes "15-love"; on the other hand, if the receiver wins the first point, the score would be "love-15." (The server's score is always given first.) After the second point is completed, the score can be either "30-love," "love-30," or "15-all." The latter case occurs if each player has won one of the two points. After the third point has been played, the possible scores are: 40-love, love-40, 30-15, or 15-30. After the fourth point has been played, if either player has won all four points, the game is over. The other possible outcomes are 40-15, 15-40, or 30-all.

If a fifth point is played, the possible outcomes are:

- if the score was 40-15:
    (a) if server wins, game is over, or
    (b) if receiver wins, score becomes 40-30.
- if the score was 15-40:
    (a) if server wins, score becomes 30-40, or
    (b) if receiver wins, game is over.
- if the score was 30-all:
    (a) if server wins, score becomes 40-30, or
    (b) if receiver wins, score becomes 30-40.

If a sixth point is played, the possible outcomes are:

- if the score was 40-30:
    - (a) if server wins, game is over, or
    - (b) if receiver wins, score is 40-40, or "deuce"
- if the score was 30-40:
    - (a) if server wins, score goes to deuce, or
    - (b) if receiver wins, game is over.

A player cannot win a game by only one point; if the score is deuce at any time, one player must win two consecutive points to complete the game. The first point after deuce is called "advantage" to the player who wins the point. If the player with the "advantage" wins the next point, the game ends; but if the other player wins the point, the score returns to deuce and play continues until a player wins two consecutive points.

Scoring the Set

A set is completed when one player has won six games, except that the winner must be at least two games ahead of the loser in order to win the set. A set may be won by scores of 6-1, 6-2, 6-3, or 6-4 games—but a score of 6-5 does not complete the set. Play must continue until one player is ahead by two games.

Scoring the Match

The winner's score in each set of a match is always reported first. For example: If player A wins two consecutive sets in a two-out-of-three match, the match score would be reported as "6-3, 6-4" if those were the scores in the two sets. If A lost one set in the match, the match score might be reported as "6-3, 2-6, 6-4"—meaning A lost the second set by a score of two games to six, but still won the match.

7.3—Baseball

▭ Numbers on First

Statistics probably play a larger part in baseball than in any other sport. They are the stuff of endless discussion—and the basis for endless debate about the performance of players, teams, and managers. Here are some of the common measures of performance:

▭ Batting Averages

A player's batting average is defined as the number of hits divided by the number of official times at bat. (Note: Walks and sacrifices do not count as official at-bats.)

 If a player came up to bat four times and struck out twice, walked once, and got one base hit, his or her batting average would be: 1 hit ÷ 3 times at bat = .333.

Ty Cobb, who played for Detroit (1905–26) and Philadelphia (1927–28), has the highest lifetime batting average in the major leagues at .367.

▭ On-Base Averages

An on-base average for a player is defined as the number of times the player reaches base divided by the number of times he or she came to the plate. (This, then, *does* include walks and sacrifices, which *aren't* considered official at-bats, but *are* considered plate appearances.)

 Using the previous example, the player's on-base average would be: 2 times on base ÷ 4 plate appearances = .500.

▭ Slugging Averages

Another closely watched statistic is the slugging average, which measures the ability of a batter to hit for power. It is defined as the total number of bases reached safely on hits, divided by the number of official times at bat.

 In five plate appearances (four official at-bats), if a player has a single (one base), double (two bases), home run (four bases), and a strikeout (no bases), his or her slugging average would be: 7 total bases ÷ 4 official at-bats = 1.750.

The all-time single-season record slugging average of .847 was set in 1920 by Babe Ruth of the New York Yankees, who also holds the lifetime slugging average of .690.

Sacrifices

A sacrifice bunt is a bunt on which the batter is put out, but a base runner advances one or more bases. A sacrifice fly is a fair or foul fly ball caught for the first or second out, but hit far enough to permit a base runner to hold his base until the ball is caught and then reach home safely. In record keeping, sacrifice bunts and sacrifice flies do not count as times at bat—and so do not affect the batting average.

Runs Batted In (RBIs)

A run batted in, or RBI, is credited to a batter when he causes a base runner to score a run. This can be done by getting a base hit, or a base on balls with the bases loaded, or by being hit by a pitch with the bases loaded, by making a sacrifice or sacrifice fly, or because of an error with fewer than two outs and a runner on third who would have scored even if the error had not occurred.

 If a player hits a grand slam (a home run with the bases loaded) and hits a single, scoring the runner on second, his or her RBI total would be five.

The all-time season record for RBIs was set in 1930 by Hack Wilson of the Chicago Cubs at 190. The lifetime RBI record is held by Hank Aaron with 2,297 RBIs.

☞ Earned Run Averages (ERAs)

A pitcher's performance is commonly measured by the earned run average, or ERA. An earned run is a run that does not result from a runner's reaching first base or scoring because of a fielding error. The ERA is the average number of earned runs given up by a pitcher during a nine-inning game. Over a season, a pitcher's ERA can be calculated as the number of earned runs charged against him or her, divided by the equivalent number of nine-inning games he or she has pitched. (Since there are three outs to each inning, the count of innings can include one-third or two-thirds fractional innings.)

 If a pitcher has pitched 135⅓ innings, or (135.33) ÷ 9 = 15.04 games, and he or she has given up 55 earned runs, his or her ERA will be 55 ÷ 15.04 = 3.65.

Bob Gibson of the St. Louis Cardinals had the lowest ERA in the modern baseball era—1.12 in 1968.

☞ On-Base Percentages

Another measure of performance for pitchers is the on base percentage for opposing batters, which is defined as the number of hits and walks divided by the number of outs plus hits and walks. (Intentional walks are not included in this calculation.)

 If this measure has a value of .235, it means that the pitcher has a 23.5% chance, on the average, of allowing an opposing batter to reach base and a 1 − .235 = .765, or 76.5% chance of retiring the average batter.

☞ Fielding Percentages

A common measure of performance for fielders is a fielding percentage, which is defined as the number of successful fielding plays divided by the number of opportunities.

EXAMPLE If a player has 200 fielding opportunities during a season and he or she has committed 10 errors, his or her success rate is 190 ÷ 200 = .950, or 95%.

Fielding Records

Fielding records also include: putouts (when a ball is caught for an out); assists (when a ball is thrown to another player for an out); errors (when a player mishandles a play he or she *should* have made); total chances (the sum of all possible fielding plays)—whether putout, assist, or error); double plays (two outs on one play); triple plays (three outs on one play); and passed balls (when a catcher mishandles a catchable pitch that the batter doesn't hit and a baserunner advances at least one base).

Batters' Records

Statistical abbreviations are quite common in baseball. Records for non-pitchers usually show the following statistics: G (games played), AB (official at-bats), R (runs scored), H (hits: singles, doubles, triples, and home runs), 2B (doubles), 3B (triples), HR (home runs), RBI (runs batted in), BB (walks, or bases on balls), SO or K (strikeouts), SB (stolen bases), CS (number of times caught stealing a base), E (fielding errors), BA (batting average), and SA (slugging average).

Pitchers' Records

For pitchers, the statistics are commonly recorded as follows: W (games won), L (games lost), ERA (earned run average), G (games appeared in), GS (games started), CG (complete games—when the starting pitcher pitches the entire game), ShO (shutouts—when the starting pitcher does not give up any runs), SV (saves—when a relief pitcher successfully comes into the game and turns back the opposing team during a rally, thereby "saving" the victory for the team), IP (number of innings pitched), H (hits allowed), R (runs allowed), ER (earned runs allowed), HR (home runs allowed), BB (walks allowed), and SO or K (strikeouts allowed).

7.4—Football

⬅ Never Fumble the Numbers

Scoring in football is based on six points for a touchdown; one point for a conversion, or point-after touchdown (PAT) (usually a kick after a touchdown; in college football, a run or pass play after a touchdown is worth two points if it succeeds); three points for a field goal (successfully kicking the ball through the uprights in the end zone); and two points for a safety (when a player is tackled in, or forced out of, his own end zone).

Statistics kept for individual players and teams largely consist of the total number of something, say, yards gained, and an accompanying average, for instance, average number of yards gained on pass receptions per game.

7.5—Basketball

✏️ Slam Dunk the Numbers

In basketball, there is only one way to score: Put the ball in the basket! But how many points you get depends on the situation. A player scores three points for a successful shot from outside the "three-point" line (the distance varies between the pros, college, and high school), two points for a "field-goal" (within the three-point line), and one point for a free throw (occurring after a player is fouled, or after a technical foul—usually a player or coach saying something offensive to the referee).

✏️ Players' Statistics

Totals and various averages are kept for the following: overall scoring, field goals, free throws, three-point field goals, assists (when you pass the ball to a teammate who, in turn, scores), rebounds (when you retrieve the ball—defensively or offensively—after an unsuccessful shot), steals, blocked shots, and fouls.

7.6—Bowling

☞ Ten-pin Bowling

Ten-pin bowling is the most popular form of bowling. Each pin you knock down counts one point. There are ten frames to a game; in each frame, each player gets two rolls. (In the tenth frame, you get a third roll if you knock down all the pins with your first two rolls.) After each frame, the pins are set up again.

STRIKES. It is a strike if you knock down all ten pins with your first roll. You get ten points plus what you make on your next two rolls. You mark a strike with an "X" in the small square within the box on the scoresheet for that frame.

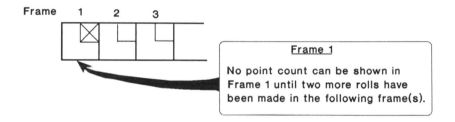

Frame   1   2   3

Frame 1

No point count can be shown in Frame 1 until two more rolls have been made in the following frame(s).

You make a perfect game—a score of 300—if you throw twelve consecutive strikes (including three in the tenth frame). You count 30 for each frame, because each strike gives you 10 plus what you make on the next two rolls. So you make $10 + 10 + 10 = 30$ in each of the ten frames, for a total of 300.

SPARES. It is a spare if you knock down ten pins in two rolls. You get ten points for the pins, plus what you make on the next roll. To mark a spare, you put a slash in the small square within the box.

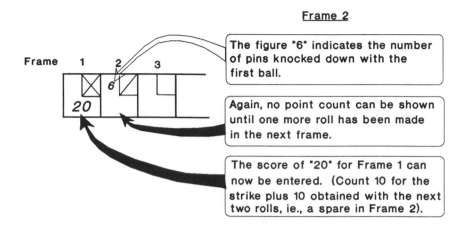

Frame 2

The figure "6" indicates the number of pins knocked down with the first ball.

Again, no point count can be shown until one more roll has been made in the next frame.

The score of "20" for Frame 1 can now be entered. (Count 10 for the strike plus 10 obtained with the next two rolls, ie., a spare in Frame 2).

OPEN FRAMES. It is an open frame if you leave one or more pins standing after rolling two balls. You get only the points for the pins you knock down.

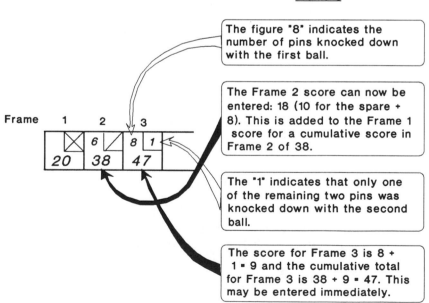

Frame 3

The figure "8" indicates the number of pins knocked down with the first ball.

The Frame 2 score can now be entered: 18 (10 for the spare + 8). This is added to the Frame 1 score for a cumulative score in Frame 2 of 38.

The "1" indicates that only one of the remaining two pins was knocked down with the second ball.

The score for Frame 3 is 8 + 1 = 9 and the cumulative total for Frame 3 is 38 + 9 = 47. This may be entered immediately.

SAMPLE SCORING. The average bowler is likely to score in the 100 to 200 range. A typical score might be:

Frame

1	2	3	4	5	6	7	8	9	10

9 /	⊠	6 2	8 1	7 /	⊠	⊠	7 /	6 2	9 / ⊠
20	38	46	55	75	102	122	138	146	166

Frame 1: (spare) 10 + 10 = 20. Total score 20.
Frame 2: (strike) 10 + 6 + 2 = 18. Total score 38.
Frame 3: 6 + 2 = 8. Total score 46.
Frame 4: 8 + 1 = 9. Total score 55.
Frame 5: (spare) 10 + 10 = 20. Total score 75.
Frame 6: (strike) 10 + 10 + 7 = 27. Total score 102.
Frame 7: (strike) 10 + 10 = 20. Total score 122.
Frame 8: (spare) 10 + 6 = 16. Total score 138.
Frame 9: 6 + 2 = 8. Total score 146.
Frame 10: (spare & strike) 10 + 10 = 20. Final score 166.

☞ Five-pin Bowling

A smaller ball is used for five-pin bowling. The setup and the values of the pins are indicated in the figure.

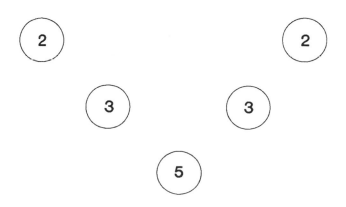

There are ten frames in a game. The pins are set up after each frame. You can roll three balls in a frame.

STRIKES. The only difference from ten-pin is that you score 15 (the total pin values) plus what you make in the next two rolls. You mark the score in the same way.

As for ten-pin, it is a perfect game if you make 12 strikes. You score 15 + 15 + 15 = 45 in each of the ten frames for a total of 450.

SPARES. The only difference from ten-pin is that you score 15 plus what you make in the next roll. If you had a strike in Frame 1 and a spare in Frame 2, your card would look like this:

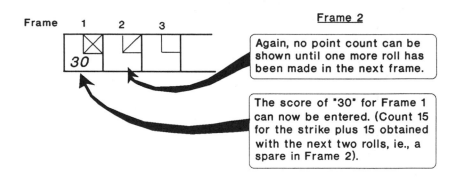

Frame

Frame 2

Again, no point count can be shown until one more roll has been made in the next frame.

The score of "30" for Frame 1 can now be entered. (Count 15 for the strike plus 15 obtained with the next two rolls, ie., a spare in Frame 2).

THE THIRD BALL. In five-pin, you can roll a third ball if there are still pins standing. In that case, the score for the frame is the total point count. There are slightly different methods of keeping score. The best is probably like this:

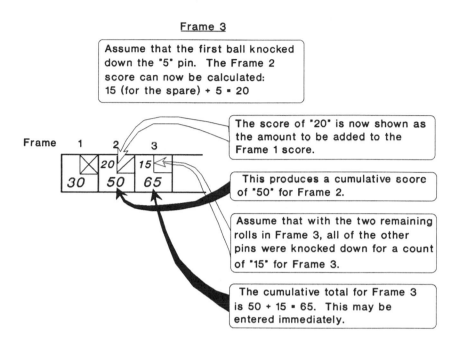

Frame 3

Assume that the first ball knocked down the "5" pin. The Frame 2 score can now be calculated:
15 (for the spare) + 5 = 20

The score of "20" is now shown as the amount to be added to the Frame 1 score.

This produces a cumulative score of "50" for Frame 2.

Assume that with the two remaining rolls in Frame 3, all of the other pins were knocked down for a count of "15" for Frame 3.

The cumulative total for Frame 3 is 50 + 15 = 65. This may be entered immediately.

SAMPLE SCORING. The average bowler is likely to score in the 150 to 275 range. A typical score card might look like this:

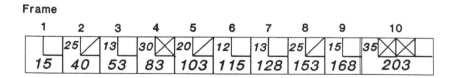

Frame

Frame 1: All pins were knocked down with three balls for a count of 15.

Frame 2: The 25 indicates 15 (for a spare) + 10 from the first ball in the next frame. Add this to 15 from Frame 1 for a total of 40.

Frame 3: The first roll counted 10, and the second roll counted 3. 13 entered + 40 = 53.

Frame 4: Count 15 (for the strike) + 15 (for the spare in Frame 5) = 30. And 53 + 30 = 83.

Frame 5: Count 15 for the spare plus 5 for the first roll in the next frame and enter 20. Add to 83 to get the total 103.

Frame 6: Score 12 for three rolls + 103 = 115.

Frame 7: Score 13 for three rolls + 115 = 128.

Frame 8: Count 15 (for the spare) + 10 (for the first ball in the next frame) = 25. Add to 128 to get the total 153.

Frame 9: Score 15 for three rolls + 153 = 168.

Frame 10: A strike with the first roll, a strike with the second roll, and five points for the third roll gives a score of 35. Add this to 168 for a final total of 203.

7.7—Hockey

☞ Team Statistics

In the National Hockey League, team standings within the divisions are determined on the basis of points, with two points for a win and one point for a tie. Team statistics are kept on wins (W), losses (L), ties (T), points (P), goals for (GF), and goals against (GA).

☞ Players' Performances

Individual players' statistics are kept for games played (GP); goals scored (G); assists (A); points (P)—one point for a goal and one for an assist; plus or minus ratio (+/−)—add one point if on the ice when your team scores, subtract one if one the ice when the opposing team scores; number of penalty minutes (PM); power-play goals (PP)—when you score with the opposing team short-handed; short-handed goals (SH)—when you score with *your* team short-handed; game-winning goals (GW)—the goal that puts your team ahead to stay; and shots on goal (SHTS). These figures can be used to calculate averages, such as average goals and average points per game.

☞ Goalies' Performances

Statistics are also kept for individual goalies' performances. These include: games played (GP); number of minutes played (MIN); average number of goals given up per game (AVG); wins-losses-ties (W-L-T); shutouts (SO); goals against (GA); shots on goal (SH); and save percentage (SV%)—number of saves divided by number of shots on goal.

8

Home, Hobbies, and Workshop Numbers

his chapter covers the numbers you need for many activities around the home, in the kitchen, at your desk, in your basement or workshop, in your garden, in your hobby room, and in your garage.

8.1—Kitchen Calculations

It used to be that working in the kitchen meant measuring, weighing, and mixing the liquid and solid ingredients called for by the recipes in the row of cookbooks in every kitchen. Today, with the rise of convenience foods and the decline of cooking from scratch, such measurements are no longer an everyday necessity in many homes. Still, standard units of measurement may be needed. Some common ones are:

- cup: volume equal to 8 fluid ounces; 2 cups = 1 pint
- fluid ounce: liquid measure; 8 fluid ounces = 1 cup = ½ pint
- gram: mass or weight—a metric unit
- ounce: weight equal to 1/16th of a pound
- tablespoon: volume equal to 3 teaspoons or 0.9 cubic inches
- teaspoon: volume equal to ⅓ tablespoon or 0.3 cubic inches

You probably still have a cookbook (or many) that will also help you if you need other numbers relating to recipes.

Other kitchen concerns include diet, nutrition, and energy values of foods. These topics have moved out of the kitchen into the areas of health and fitness, which we address in Chapter 4, particularly in Section 4.8, page 83.

Ways to keep track of your electricity or gas consumption in the kitchen are addressed in the general discussion of these topics in Section 8.6, page 182, and in Section 8.7, page 186.

Bar Codes

The bar code symbol you see on most of your grocery items is composed of a set of numbers that tells the cash register in the store what kind of product it is, who made it, and something about its size, color, and the like. It does not say what the price is; that has to be programmed into the cash register separately.

Basically, the code consists of eleven numbers that are made up of groups of 0's and 1's. The scanner crosses the black and white bars of the symbol and reads a dark strip as a 0, and a white strip as a 1. (A bar can be thin, if it has only one strip, or thicker if it has two or more strips side by side.)

In addition to the coded numbers, there are edge markers, a center marker, and an extra digit that is used as a check that will indicate whether the numbers make sense or contain an error of some kind.

The system is administered by the Uniform Code Council in Dayton, Ohio. It began in 1973 and is now used on products other than grocery items, such as magazines, compact discs, and tags attached to clothing. And look on this book's back cover!

8.2—Fabric for Your Pattern

The number of yards of fabric you will need for a particular pattern is usually marked on the pattern, but if it is not marked and you have to estimate it, you can have difficulty because it can vary greatly depending on the width of the fabric. Table 8.2 can help you make your estimate if you can measure how many yards you need at one width and you want to convert this to how many you need at another width.

TABLE 8.2

Yards of Fabric You Need

FABRIC WIDTH IN INCHES

	32	35–36	41	44–45	52–54	58–60
	1⅞	1¾	1½	1⅜	1⅛	1
	2¼	2	1¾	1⅝	1⅜	1¼
Y	2½	2¼	2	1¾	1⅝	1⅜
	2¾	2½	2¼	2⅛	1¾	1⅝
A	3⅛	2⅞	2½	2¼	2¼	1¾
	3⅜	3⅛	2¾	2½	2¼	1⅞
R	3¾	3⅜	2⅞	2¾	2¼	2
	4	3¾	3⅛	2⅞	2⅜	2¼
D	4⅜	4¼	3⅜	3⅛	2⅝	2⅜
	4⅝	4½	3⅝	3⅜	2¾	2⅝
S	5	4¾	4	3⅝	2⅞	2¾
	5¼	5	4⅛	3⅞	3⅛	2⅞

Source: Rutgers Cooperative Extension Service, State University of New Jersey.

8.3—The Home Office

If you sometimes use a desk at home for office work, correspondence, and other paperwork, here are some numbers you might need.

✏ U.S. Postage Rates

TABLE 8.3

U.S. Postage Rates (as of the end of 1991)

First Class mail:

Weight	Rate	Weight	Rate
1 oz	$0.29	7 oz	$1.67
2 oz	0.52	8 oz	1.90
3 oz	0.75	9 oz	2.13
4 oz	0.98	10 oz	2.36
5 oz	1.21	11 oz	2.59
6 oz	1.44		

Post Cards: $0.19 each (Maximum size: 4.25 × 6 in.,
Minimum size: 3.5 × 5 in.)
Consult your postmaster for rates on Priority, Express, Third Class, and Fourth Class mail.

(Source: U.S. Postal Service)

✏ ZIP Codes

The first digit in the 5-digit ZIP Code designates one of ten large areas numbered from 0 in the Northeast to 9 in the West. Each area covers three or more states or territories. The areas are divided into sub-areas, usually about ten, with a sectional center in each. The second and third digits of the code, with the first, identify the area and sub-area. The last two digits identify the town or local post office within the sub-area.

✏ ZIP + 4

Four more numbers have been added to ZIP Codes to further pinpoint addresses. The first two numbers identify a group of streets or buildings, and the second two identify a group of boxes or, for example, a department within a firm.

▱ Types of Paper

Paper can be classified in terms of pounds:

- 20-pound is standard typing paper
- 24-pound is standard letterhead paper
- 65-pound is for postcards or business cards

The weight figure comes from the weight of a ream of paper, which usually consists of 500 sheets.

▱ Lead Pencils

The hardness of the lead (actually, graphite) in a drawing pencil is classified by a peculiar letter and number scale: 4B, 3B, 2B, B, HB, F, H, 2H, 3H, 4H, etc. Here B is soft, H is hard, and F is fine. The B numbers get softer to the left, and the H numbers get harder to the right in this scale.

Everyday pencils are usually classified as 2, 2.5, or 3, with the lead softer and darker at the lower end of the numeric scale, and harder and lighter at the higher end.

▱ Type Size

Type size is another esoteric area that is understandable only by the specialist. Roughly speaking, points measure heights and picas measure widths; there are 72 points or 6 picas in an inch. The "pica" that is standard on many typewriters has characters that are one pica wide and twelve points tall, and there are six lines per inch. With personal computers, of course, you frequently have a large selection of typefaces and point sizes from which to choose, depending on the capabilities of your printer.

8.4—Wallpapering

➣ Off-the-Wall Numbers

Before you buy wallpaper, you will want to estimate the number of rolls you will need and how much it might cost you.

The terms used to describe wallpaper can be confusing. The price is often quoted as the "cost per single roll," although it is sold only in double rolls. So you may need to double the quoted price.

Most double rolls sold today contain between 56 and 70 square feet of paper. The nominal dimensions of the most common rolls are 20.3 inches by 41.4 feet = 70 square feet.

➣ Allow for Wastage

In estimating the number of rolls you need, you must allow for wastage—for trimming at the ceiling and floor, and for pattern-matching where necessary. If the pattern repeats itself every 20 inches, which is typical, there may be something like 20% wastage. In addition, you will have wastage owing to cut-outs for windows, doors, and arches. (Although you don't paper over these openings, they cause wastage because you have to use full-length strips at their edges even though the full width of the paper may not be used. Also, you will have wastage above and below the openings if you must match a pattern.)

➣ Estimate What You Will Need

To make a rough estimate of the number of rolls you need, measure the ceiling height and calculate the length of a single strip, allowing for trim at the top and bottom and for pattern-matching. After finding this strip length, you should divide it into the total length of the double roll.

EXAMPLE Assuming that the ceiling height is 8½ feet, and allowing 16 inches for trim and pattern-matching, you have a strip length of 118 inches. When this is divided into the total double-roll length of 41.4 feet, you obtain 4.2 for the number of strips in a double roll. This would be reduced to 4 because the fraction will be wasted. You would note that the width of a strip is 20.3 inches, and that 4 strips will stretch across a distance of 4 × 20.3 = 81.2 inches. If the total distance around the room is, say, 40 feet, then by dividing you find (40 x 12) ÷ 81.2 = 5.9 double rolls would be needed to stretch around the room. Because you have not subtracted for any openings, this would be the maximum number of rolls needed.

A rule of thumb: Subtracting one double roll for every four doors, windows, etc., gives a satisfactory estimate. So you arrive at an estimate of six double rolls, less one for every four doors, windows, and the like.

You can check your estimate by using Table 8.4 or you can get a table from a dealer.

TABLE 8.4

Rolls of Wallpaper Required

HEIGHT OF THE ROOM	DISTANCE AROUND THE ROOM							
	Feet	20	36	40	44	48	56	60	72
	NUMBER OF DOUBLE ROLLS NEEDED							
7.5-8.5 feet		3	5	6	6	7	8	8	10
8.5-9.5 feet		3	6	7	7	8	9	9	11

(For each four doors, windows, etc., subtract one double roll.)

In the previous example, in which the distance around the room was 40 feet and the room height was 8.5 feet, you can see that the table shows that you'd need six (or possibly seven) standard double rolls. This confirms the previous estimate, but it indicates that you should use the higher estimate as long as unused rolls are returnable. If you don't get enough at the start, it may be hard to get the exact pattern number and lot number later—in which case you run the risk that the colors won't match exactly.

8.5—Painting

✏ Paint by Numbers

How much paint to buy and what it will cost are the only numerical questions in a painting project.

In addition, though, you should minimize the amount you buy, because unused paint is a major component of today's toxic-waste problem. On the other hand, if you are using a special color mix, you have to be sure you get enough, because color matching may also be a problem. So you have a balancing act to perform, and you need to look at the numbers to make your decision.

✏ Area to Cover

The area to cover is the first thing to work out. This will usually be a simple problem of measuring length and width and multiplying to get area in square feet. You can do this for all the rectangles involved (including the long and narrow rectangles for the trim) and then add the results. If there are unusual shapes, such as circles or cylinders, you may need to look for a formula in A.8, pages 261 to 267.

✏ Coverage Factor

For the coverage factor of the paint, you will have to look at the label on a can. It will tell you how many square feet a quart or a gallon will cover for different surfaces, such as new wood and previously painted wood.

✏ Price Is Another Factor

These numbers will let you work out how much paint to buy, avoiding buying too much or too little. Since paint is cheaper by the gallon than by the quart, you will have to work out the prices of your options and make your choice.

8.6—Electrical Power in the Home

➡ Current Numbers

Most of the numbers concerning electric current in the home can be related to a simple equation called Ohm's Law, which says that the current that will flow through an electrical device equals the voltage applied divided by the resistance in the circuit. The units normally used are amperes (amps), volts, and ohms, so the equation says: Amps = Volts ÷ Ohms.

 EXAMPLE **If a toaster is plugged in to a 110-volt circuit, and its resistance is 10 ohms, the current is 110 volts ÷ 10 ohms = 11 amps.**

➡ Fuses and Circuit Breakers

Household circuits are protected against excessive current flow by fuses or circuit breakers. These are rated and marked according to the maximum current allowed to flow in the circuit before they act. They will be labeled 15-amp, 20-amp, 30-amp, etc. You should use 15-amp fuses to protect most household circuits.

 EXAMPLE **If a toaster drawing 11 amps is plugged into a circuit, and at the same time an electric iron drawing 13 amps is plugged into the same circuit, the total current would be 11 + 13 = 24 amps. Clearly, if the circuit is protected by a 15-amp fuse, the fuse would blow and the circuit would be opened to prevent a possible fire caused by overheating.**

➡ Watt Numbers?

Electrical power is important because you pay for the power you use in units of watts or thousands of watts (kilowatts). Most of the numbers concerning power can be related to an equation that says that power equals current multiplied by voltage. When this is combined with Ohm's Law, you have two relationships that can be useful in answering questions about power:

1: watts = amps × amps × ohms

2: watts = (volts × volts) ÷ ohms

 EXAMPLE If the toaster in the previous example has a resistance of 10 ohms and the voltage is 110 volts, the power equation says: power = (110 × 110) ÷ 10 = 1,210 watts, or 1.21 kilowatts.

⇨ Cost of Electricity

Electric companies usually bill you for the amount of power you use during a one-month period. The amount used is measured in kilowatt hours, that is, the power in kilowatts multiplied by the number of hours that the power is being used.

 EXAMPLE If a 100-watt bulb burns continuously for 24 hours, the amount of electricity used can be found by multiplying the power, 0.1 kilowatts, by 24 hours to obtain 2.4 kilowatt-hours.

⇨ Rates Vary with Use

Electric companies charge one rate for the first X number of kilowatt-hours and a lower rate for additional consumption. Typically, you might be charged 12 cents per kilowatt-hour for the first 250 kilowatts and 10 cents for any more you use. Table 8.6 gives average figures for the cost of operating various appliances for typical periods during a month. (It assumes the unit cost is 10 cents per kilowatt-hour.)

TABLE 8.6

Monthly Consumption and Cost

APPLIANCE	KILO-WATTS	HOURS USED	KILO-WATT-HOURS	COST, $ PER MONTH
Water Heater	4.5	90	405	40.50
Stove	12.2	30	366	36.60
Refrigerator	0.6	240	144	14.40
Air conditioner	1.5	90	135	13.50
Freezer	0.34	240	82	8.20
Dryer	4.8	12	58	5.80
Color TV	0.33	120	40	4.00
Dishwasher	1.2	30	36	3.60
Iron	1.2	10	12	1.20
100-Watt Bulb	0.1	120	12	1.20
60-Watt Bulb	0.06	120	7	0.70
Washing Machine	0.5	12	6	0.60
Coffee Pot	0.9	6	6	0.60
Toaster	1.2	2	2	0.20

If you use electrical heating in your home, look at the numbers in Section 8.8 on Home Heating, page 187.

⬚ Battery Systems

Boats, travel trailers, and vacation homes may have battery-powered electrical systems with backup generators to charge the batteries. The power drain on such systems can be a matter of concern. You may have to add up the current drains of the appliances you have before adding a new load on the system.

If you want to add an appliance such as a refrigerator to a 12-volt system in a travel trailer, you will need to check the loading on the system to ensure that you don't exceed the 15-amp fusing on the circuit. The equation Power = Voltage × Current can be reversed to Current = Power ÷ Voltage—a useful step in checking current drains.

The existing load on a one-circuit system with a 15-amp fuse might be:

Interior lighting: 4   12-watt bulbs; current = 48 ÷ 12 = 4 amps

Outside lighting: 12   3-watt bulbs; current = 36 ÷ 12 = 3 amps

Radio: 12 watts; current = 12 ÷ 12 = 1 amps

Television: 24 watts; current = 24 ÷ 12 = 2 amps

Water pump: 36 watts; current = 36 ÷ 12 = 3 amps

Total current with existing load = 13 amps

Therefore, if you should want to install a 100-watt refrigerator, drawing 100 ÷ 12 = 8.3 amps, you would have to add a second circuit, fused for 15 amps, to handle the added load.

8.7—Natural Gas in the Home

➯ Reading Your Gas Meter

If you use natural gas in your home, your gas meter will measure how much you use. The dials on your meter are set to read hundreds of cubic feet.

 EXAMPLE **If the four dials read 6,543 and they read 5,432 last month, your consumption for the month was 6,543 − 5,432 = 1,111 × 100 cubic feet, or 111,100 cubic feet.**

These units are used because 100 cubic feet of natural gas delivers approximately 100,000 British Thermal Units (BTUs)—which, for conciseness, is defined as one therm, or quantity of heat. There is some variation in the heating qualities of natural gas; your bill shows the relationship between the meter-reading and the number of therms delivered.

➯ Typical Therms per Hour

Average consumption rates, in therms per hour of operation, are as follows: oven (1.3); burner on a stove (1.2); space heater (0.8); and pool heater (2.5). Your washer and dryer will each use about 0.1 to 0.3 therms per load. Your gas bill will show you the price per therm that you are paying. (See Section 8.8, page 187, for a discussion of home heating.)

8.8—Home Heating

✑ Year-to-year Cost Comparisons

If you want to compare the cost of heating your home from year to year, you will need to keep detailed records of your fuel consumption. You may, for example, want to see what you are saving by improving the insulation in your home, or by lowering the setting on your thermostat, or by some other action. The problem is that the year-to-year average temperature is different, and you must allow for this variation. You don't have to keep climatic records yourself, because they are available from your nearest weather office.

✑ Heating-Degree-Days

The units used are "heating-degree-days." Each day, the mean of the maximum and minimum temperature is found, and the difference between this mean and a standard temperature of 65 degrees Fahrenheit is calculated. This difference is the number of "heating-degrees" for that day. At the end of the month, the total is found by adding the daily values.

Figure 8.8 shows a typical plot of monthly totals for a year.

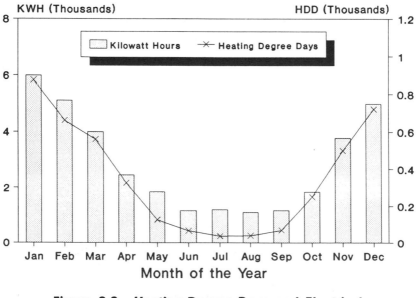

Figure 8.8—Heating-Degree-Days and Electrical Consumption Over a Year

Also shown in Figure 8.8 is a plot of the consumption of electricity in kilowatt-hours. You might have kept records by reading your meter at the end of each month and adjusting the reading by subtracting your estimate of the consumption that was due to other uses, using values like those in Table 8.6, page 184.

Heating-degree-days are useful numbers for this kind of analysis, because fuel consumption varies in an approximately linear fashion with these units—that is, if heating-degree-days are doubled, generally fuel consumption will double, and this will be true whether the doubling occurs over a short time or over a longer period.

Heating with Oil or Gas

You can draw a graph similar to Figure 8.8 if you heat your home with either oil or gas. You can easily read your gas consumption for the month easily from the meter. Oil consumption is more difficult to determine, because the gauges aren't as accurate. In that case, you may need to arrange for regular refilling of your tank by your supplier, preferably when it isn't as cold—prices are highest in the dead of winter.

Heating by Firewood

Conventional open fireplaces are extremely inefficient because they lose most of their heat up the chimney, but many more-efficient wood-burning units have come into use in recent years.

It is difficult to keep precise monthly records of consumption of firewood, but you can at least determine your annual cost and adjust for the total heating-degree-days over the year. The quantity supplied is measured by volume, usually in terms of a "cord." (See B.1, page 292.) It is important that firewood be dry rather than green, because the heating value of dry wood can be as much as 40% higher than that of green wood.

The heating value of equal weights of dry wood is practically the same for all species of tree, but because some species have a higher density than others, the heating value of equal volumes of different species can vary by as much as 100%.

The units used to measure the heating value per unit volume of dry wood are British Thermal Units (BTUs) per standard cord (See B.1, page 291.)

The common species can be put into four groups with similar heating values, as shown in Table 8.8A.

TABLE 8.8A

Heating Value of Common Firewoods

Group 1	Group 2	Group 3	Group 4
15 to 17 million BTU/cord	18 to 22 million BTU/cord	23 to 25 million BTU/cord	26 to 32 million BTU/cord
balsam fir	trembling aspen	white birch	yellow birch
white spruce	hemlock	black cherry	red oak
white cedar	large tooth aspen	tamarack	beech
basswood	manitoba maple	red maple	sugar maple
white pine	silver maple	white elm	bitternut hickory
balsam poplar	green ash	white ash	white oak
butternut	black ash	red elm	shagbark hickory
			rock elm

SOURCE: Ontario Ministry of Natural Resources

▭ Reducing Heat Loss

In recent years, more and more homeowners have sought to reduce heat loss from their homes by improving their insulation. The numbers that apply in this field are R-values—meaning resistance to the flow of heat. The R-value of a material depends on its type and its thickness. For instance: A 4-inch-thick blanket of fiberglass with an R-value of 3 per inch will have a net R-value of $4 \times 3 = 12$.

There are some variations in recommended R-values for different regions of the country, but, in general, minimum values shown in Table 8.8B can be used.

TABLE 8.8B

Recommended R-values

For existing exterior walls R-11
For new-construction exterior walls R-19
For floors over basements and crawl spaces R-19
For ceilings under ventilated attics R-38

Source: U.S. Department of Energy

If you want to be more precise, you can ask a local insulation supplier in which Insulation Zone of the country you live. The U.S. Department of Energy has drawn up tables that are keyed to the first three digits of the ZIP code. (See 8.3, page 176.)

Table 8.8C shows the thickness of materials needed to provide the recommended R-values.

TABLE 8.8C

Inches of Insulation Needed

R-VALUE	BLANKETS	LOOSE FILL	
	FIBERGLASS	FIBERGLASS	CELLULITE
R-11	3.5–4	5	3
R-13	4	6	3.5
R-19	6–6.5	8–9	5
R-22	6.5	10	6
R-26	8	12	7–7.5
R-30	9.5–10.5	13–14	8

SOURCE: U.S. Department of Energy

8.9—Construction Around the House

Planning Numbers

Almost all construction projects at home create numerical needs—for estimating the materials you will need and their costs. Often, the first steps will be to draw a rough-sketch plan and to make a list of the materials you will need.

Carpentry

In general, you will want to draw a design or obtain a prepared plan, with material requirements, for any carpentry project. In most cases, the procedure will be to make careful measurements and work out a detailed parts list—with the help of a dealer, if required. Catalogs from lumber and hardware dealers can also be useful, since they provide specifications and prices with which you can draw up your materials list and estimate the total cost of your project.

Lumber Numbers

Grades of lumber are stated in words more often than in numbers, though they can have numbers attached as well. In general, grades indicate the quality of the lumber in terms of the size and number of knots and other defects. Since different kinds of wood have different grading systems, you will need to discuss with a lumber dealer the grade you require.

The measurements that describe lumber are the sizes before finishing. By the time the rough lumber reaches the yard where you buy it, it has been trimmed in thickness and width. Thickness is reduced by about one-quarter inch for one-inch lumber and about one-half inch for other size lumber. Width is reduced by about one-half inch. Therefore, a piece of 2 × 6 lumber actually measures about 1½ inches by about 5½ inches.

You will find lumber sold either by "linear feet" or by "board feet" (BF). Linear means the length measured in feet. One BF measures one foot long, one foot wide, and one inch thick. Therefore, one linear foot of 2 × 12 lumber measures 2 BF because it is two inches thick. To calculate the BF in a board, you need to calculate its area in square feet and multiply this area by its thickness in inches.

 EXAMPLE A twelve-foot piece of 2 × 4 lumber has 12 × (4/12) = 4 square feet area. Multiply by two inches and you have the answer—8 BF.

✏ Concrete Construction

In any home project involving concrete, you should be sure to mix the cement, sand, coarse aggregate, and water in the right proportions.

Cement is usually sold in sacks of one cubic foot volume, weighing 94 pounds. For most backyard projects, you should add from four-and-one-half to six gallons of water to each sack of cement, depending on how wet or dry your sand is. You should add two-and-one-half cubic feet of sand and three-and-one-half cubic feet of coarse aggregate (gravel, for example) to the mix. In these proportions, the amount of materials needed for each cubic yard of concrete will be six sacks of cement, fifteen cubic feet of sand, and twenty-one cubic feet of coarse aggregate.

When you know how many cubic yards of finished concrete you need, you can use the "three-halves" rule of thumb to estimate how much cement, sand, and aggregate (in the approximate proportions of 1:2:3) you will need. The rule is that the combined ingredients at the start are three-halves of the final volume of the mix. (The sand fills spaces between aggregate particles, and the cement fills spaces between sand particles.)

 EXAMPLE To make a wall ten feet by four feet by six inches, you need 10 × 4 × .5 = 20 cubic feet. Applying the three-halves rule gives 20 × 1.5 = 30 cubic feet, made up of:

cement: 1/6 of 30 = 5 cubic feet

sand: 2/6 of 30 = 10 cubic feet

aggregate: 3/6 of 30 = 15 cubic feet

If you want to use concrete blocks, you can estimate how many you need from the rule of thumb that 113 blocks make 100 square feet of wall (using standard 8 × 16 × 4 or 8 inch blocks). You need about six cubic feet of mortar for 100 square feet of wall.

Brick Construction

If you are using bricks in a construction project, you can estimate how many you will need using the figures in Table 8.9. (The dimensions of standard bricks are 2½ × 3 3/4 × 8 inches.)

TABLE 8.9

Bricks per Square Foot of Wall

JOINT THICKNESS	NUMBER OF BRICKS	JOINT THICKNESS	NUMBER OF BRICKS
¼ in.	7.0	⅝ in.	5.80
¾ in.	6.55	¾ in.	5.50
½ in.	6.15		

Ceramic Tiles

If you are applying ceramic tiles to a wall, you can estimate how many you will need using the rough figure of eight tiles per square foot of surface. (Standard ceramic tiles are four-and-one-quarter inches square.) It is important to obtain some spare tiles because the color of tiles, even from the same lot and company, can be noticeably different.

Floor Tiles

If you are laying floor tiles, or carpet tiles, to cover 100 square feet, you will need 180 9-by-9-inch tiles or 100 12-by-12-inch tiles. To allow for cutting and wastage, you should add about 10%.

One method of laying tiles in a rectangular area is to start at the center of the space, laying two strips at right angles across the length and width of the space. Then fill in the four areas. In this way, partial tiles will be needed only around the edges of the space.

8.10—Gardening

▭ Numbers with Green Thumbs

Most of the calculations you must make as a gardener are concerned with determining areas and volumes, mixing solids or liquids in certain proportions, spreading materials in certain concentrations, adjusting the acidity of your soil, and providing fertilizer for the types of plants you want to grow.

For projects involving laying out beds, pathways, patios, building fences, and the like, you will need to draw some kind of plan and calculate areas and volumes. Most calculations will be simple, though you will have to be careful about the units of measurement you are using. Usually, package labels will guide you on the amounts of material you will need—for example, in estimating how much peat moss, compost, or lime you need to cover the area you are treating.

You may need to calculate the quantity of liquid you'll need for purposes such as fertilizing or controlling pests. Again reading package labels will help here.

Fertilizers are labeled according to the percentages by weight of the three principal active ingredients: nitrogen (N), phosphorous (P), and potassium (K).

 A general-purpose fertilizer, suitable for most garden plants and lawns, might be labeled 20-4-10, indicating that it contains 20% nitrogen, 4% phosphorous, and 10% potassium. In this case, the remaining 66% of the material is inactive.

The proportions of these ingredients, rather than the absolute amounts, are what's important. So a 10-10-10 fertilizer is proportionally less nutritive than a 20-20-20 fertilizer—the difference between the two is that the 10-10-10 is 70% inactive, while the 20-20-20 is 40% inactive.

In general the choice of the best fertilizer for your needs will depend, for instance, on the soil you have, the plants you are growing, or whether you want to encourage leaf, stem, or root growth.

⬭ The pH Level of Soils

As explained in Section 5.9, page 113, the acidity or alkalinity of substances can be expressed in terms of a scale that runs from 0 to 14. At a pH level of 7.0, the midpoint of this scale, the substance is neutral—that is, neither acidic nor alkaline. Values less than 7 indicate acidity, and values above 7 indicate alkalinity. The scale is logarithmic, which means that a decrease of one pH value indicates a tenfold increase in acidity. Anything less than pH 6 is said to be "acid soil." Most plants grow best in soils within the range of pH 5.5 to pH 7.2; a pH level of 6.5 is probably optimum for most flower and vegetable gardens. You'll need expert advice to obtain the best results for particular conditions; it may be necessary to test your soil and to adjust its pH level accordingly.

8.11—Radio and Television Frequencies

⟐ **Numbers in the Ether**

Radio and television broadcasts come to your home via electromagnetic waves in space. Their frequencies are near the low end of the vast spectrum of electromagnetic frequencies, including direct current from a battery; electrical-power frequencies, such as the 60-cycles-per-second power you have in your home; microwave frequencies used in ovens; visible light frequencies; X-rays; and nuclear and cosmic rays.

The units used to express these frequencies are hertz (hz), the assigned name for cycles per second; kilohertz (khz), or thousands of hertz; and megahertz (mhz), or millions of hertz. ("Hertz" is named after Heinrich Rudolf Hertz [1857–94], a German physicist who studied electromagnetic waves.)

Table 8.11 shows the range of frequencies.

TABLE 8.11

Electromagnetic Wave Frequencies

FREQUENCY	APPLICATIONS
0 hz.	Direct current
60 hz.	Alternating current
550–1600 khz.	AM radio stations
54–88 mhz.	TV channels 2 to 13
88–108 mhz.	FM radio stations
460–470 mhz.	CB radio
470–890 mhz.	TV channels 14 to 83
10^{10} hz.	Microwaves
10^{15} hz.	Visible light
10^{18} hz.	X-rays
10^{21} hz.	Gamma rays
10^{25} hz.	Cosmic radiation

AM radio stations are each assigned a 10 khz. band width; FM stations have a 200 khz. band width; and TV stations have a 6,000 khz. band width.

8.12—Camera Settings

✍ Numbers on Tripods

Most of today's cameras are largely automatic, freeing the photographer from the need to calculate the best combination of shutter speed and lens opening for each picture. Sometimes, however, you may wish to use the manual settings of your camera to achieve a particular effect—in which case you will need to understand the numbers relating to shutter speeds, f stops and depths of field.

✍ Shutter Speeds

The shutter speeds listed on your camera indicate fractions of a second; 500 on the shutter-speed dial means that the shutter of the camera is open for 1/500th of a second. The scale on a camera typically reads 500, 250, 125, 60, 30, 16, and B for "bulb" (or X for "flash"). For normal sunny conditions, you might use a shutter speed of 1/125—with shorter exposures for brighter conditions and longer exposures for cloudy and darker conditions. When the shutter stays open longer than 1/125th of a second—1/60th, 1/30th, or 1/16th of a second—any movement of the camera will likely blur the picture. To prevent blurring, you must use a tripod or other stabilizing device. To "freeze" fast-moving objects, you must use short time exposures—for example, 1/250th or 1/500th of a second.

✍ f Stops

To let more (or less) light through the shutter in the short time it is open, you enlarge (or reduce) the lens opening by varying your camera's f stop setting. A special scale is used for this setting, which differs depending on the lens. A representative set of f stop settings is: $f/2$, $f/2.8$, $f/4$, $f/5.6$, $f/8$, $f/11$, and $f/16$.

These numbers are actually ratios that are found by dividing the focal length of the lens (that is, the distance from the lens to the film) by the diameter of the lens opening.

EXAMPLE For a lens with a focal length of 240 millimeters and a lens-opening diameter of 30 millimeters, the *f* stop would be 240 ÷ 30 = 8, while for a lens-opening diameter of 15 millimeters, the *f* stop would be 240 ÷ 15 = 16.

Note that each of the preceding photographic scales is an inverted scale. The larger the number representing the shutter speed, the shorter the exposure time; the larger the *f* stop number, the smaller the lens opening.

Depth of Field

Because of the properties of lenses in general, more of the scene is in sharp focus the higher the *f* stop number. In other words: The "depth of field" increases at higher *f* stops—which means that objects close to the camera can be in focus at the same time as objects farther away. For this reason, it is often desirable to use high *f* stop numbers (small lens openings), which in turn means that you must use longer exposure times (slower shutter speeds) to be sure that enough light gets through to form a picture.

Choose the Best Combination

In fact, there are many combinations of shutter speed and lens opening that will result in the same exposure of the film. You must choose the combination that is best for a particular picture at a particular time. Some factors that you should consider are: the amount of light available, the need to stop any action in the scene, the depth of field you want, and the availability of a tripod. If you have an exposure meter, you can use it while trying different combinations before you take the picture.

8.13—Binoculars and Telescopes

➪ Binocular Numbers

Binoculars have two identifying numbers. The first represents the power (or magnification) of the binoculars. For example: 6× indicates that the glasses magnify six times, which means that they make an object appear six times as large as it would appear without the glasses.

The second figure gives the diameter of the front lenses (the "objective" lenses) in millimeters. For example: For a pair of binoculars with 30 millimeter lenses with 6× magnification, the numbers would be 6 × 30.

Binoculars' light-gathering capability—that is, their brightness—is expressed by dividing the lens diameter by the magnification and squaring the result.

 EXAMPLE For 6 × 30 binoculars, the relative brightness is (30 ÷ 6) × (30 ÷ 6) = 25; for 8 × 50 binoculars, it is (50 ÷ 8) × (50 ÷ 8) = 39. These figures mean that the 8 × 50 pair collects much more light than the 6 × 30 pair, so distant objects can be seen more clearly.

Unfortunately, glasses with a higher magnifying power also have a smaller field of view, which means that the area that can be viewed at one time is reduced. Larger binoculars are more difficult to hold steady on the desired object, both because of their smaller field of view and also because they are larger and heavier.

➪ Telescopic Numbers

When you look at a distant object through a telescope, light from the object passes through an objective lens or reflects from an objective mirror to form a first image at a distance from the lens or mirror called the focal distance. There is then an eyepiece lens through which you see a magnified second image at a different focal distance from the eyepiece lens.

The three "powers" associated with a telescope are light-gathering power, resolving power and magnifying power.

➱ Light-gathering Power

A telescope concentrates light from a wide beam into a small beam that can enter your eye, and in this way the telescope increases the apparent brightness of the object. Its light-gathering power depends on the area of the objective, and therefore is proportional to the square of the diameter of the objective.

 EXAMPLE

A dark-adapted eye has a diameter of about one-quarter inch; a telescope with a two-inch objective has a diameter eight times as large— and, therefore, admits 8 × 8 = 64 times as much light as the unaided eye.

➱ Celestial Magnitude

The apparent brightness of celestial objects is expressed by a scale of magnitude. On this scale, the unaided eye can see stars of Magnitude 6. Magnitude 7 applies to stars that are fainter by a factor of 2.5, and Magnitude 8 applies to stars that are fainter by $(2.5)^2$, and so on. Therefore, in the example, we can say that when 2.5 is raised to an unknown power ×, the result is 64—from which we conclude that × = 4.5, and therefore we find that a two-inch telescope allows stars of Magnitude 6 + 4.5 = 10.5 to be seen.

➱ Resolving Power

A telescope allows closely spaced distant objects to be distinguished from one another as separate objects. Light from distant objects comes into the telescope at slightly different angles; the closer together the objects are, the smaller the angular difference will be. The smallest angular differences that can be separated by different-sized telescopes are shown in Table 8.13.

TABLE 8.13

Resolving Power of Telescopes

LENS DIAMETER IN INCHES	4	5	6	8	10	12
ANGLE IN SEC. OF ARC	1.25	1	0.8	0.6	0.5	0.4

✏ Magnifying Power

Telescopes have magnifying powers like those of binoculars; in the same way, higher magnifying power means that the field of view is reduced in proportion. The magnifying power of a telescope is equal to the ratio of the focal length of the eyepiece lens to the focal length of the objective lens.

8.14—Microscopes

☞ Microscopic Numbers

The optical system of a microscope works like a telescope's, with an objective lens and an eyepiece. Magnifying powers and resolving powers are similar. For standard microscopes, using visible light, the highest magnifications possible are in the 500-times range when the full resolving power of the instrument is used.

☞ Shorter Wavelengths

To achieve higher resolving powers, one must reduce the wavelength of the radiation; for this reason, ultraviolet light is sometimes used. In that case, the optical elements must be quartz instead of glass to reduce absorption. To obtain further increases in resolving power, one must use beams of electrons, which can have wavelengths several thousand times shorter than visible light. Electron microscopes can resolve objects separated by one millionth of a centimeter or less, and their magnifying powers can be about 250,000 times.

8.15—Car Facts and Figures

We analyzed the costs of buying and operating a car in 1.2, page 6. In this section, we cover several other automotive numbers.

▷ Engine Displacement

The combined volume of the cylinders of a car's engine is used as a measure of the size of the engine, which is an indicator of the engine's power and output capability.

 EXAMPLE **If an engine has four cylinders, each having a volume of 30.5 cubic inches, the total volume in which fuel is burned is 122 cubic inches.**

The volume of a cylinder can be expressed by the formula in A.8, page 266. It depends on the "bore" (diameter) of the cylinder and on the "stroke" (the distance that the piston travels up or down in each cylinder).

▷ Engine Power

The power of a car's engine depends on its displacement and other factors. The original measure of one horsepower was the power required to raise a 550-pound weight a distance of one foot in one second—that is, 550 foot-pounds per second. (See B.1, page 293.)

Modern passenger-car engines range from about 80 to 300 horsepower. For comparison: A motorcycle's power might be in the range of ten to one hundred horsepower, a lawnmower's power might be about three horsepower, and a light aircraft's power might be about thirty horsepower.

▷ Mileage and Fuel Consumption

Table 8.15 gives some representative figures for mileage and fuel consumption of modern cars sold in North America.

TABLE 8.15

Mileage and Fuel Consumption

TYPE OF CAR	ENGINE DISPLACEMENT IN CUBIC INCHES	CYL-INDERS	TRANS-MISSION	MILES PER GALLON CITY/HIGHWAY	
LUXURY	305	8	AUTO	16	24
FULL SIZE	183	6	AUTO	18	29
MID SIZE	122	4	AUTO	24	34
COMPACT	91	4	MANUAL	29	39
SUB-COMPACT	61	3	MANUAL	39	47

✏ Octane Rating

The grading of gasolines at the pump is done on the basis of an "octane rating." Typically, premium gas has a rating of about 91 and regular gas has a rating of about 87.

Octane, one of the ingredients of crude oil, can contribute to the prevention of "knocking" caused by premature ignition of the fuel in your car's engine. An octane rating of 91 indicates that the gasoline has the anti-knock performance of a mixture of 91% octane and 9% of a lower-performance ingredient. The recommended procedure is to choose the lowest-octane gasoline that will provide good performance in your car without knocking. Using a higher grade of octane than necessary is wasteful and unnecessary.

✏ Oil Grade

For multigrade oils, the numbers indicate the "viscosity," or ability to flow, at cold temperatures (marked W for Winter) and at warm temperatures. The first number in the grade designator may be 5W, 10W, 15W, or 20W, which are arbitrarily assigned numbers obtained from measurements of a sample at 0 degrees Fahrenheit. Grade 5W flows through a small hole in a certain time; 10W takes twice as long; 15W, four times as long; and 20W, eight times as long.

The second number gives the measured results at 210 degrees Fahrenheit. This is called the "regular" rating; it is the only number given for single-grade oils. It will be 20, 30, 40, or 50, an arbitrary number based on the time the oil takes to flow through a small hole at 210 degrees Fahrenheit.

Your operator's manual will tell you the grades of oil you should use for different conditions. In general, the basis for choosing the correct oil is to use an oil with a low W rating for low temperatures and a high regular rating for high temperatures, or use an oil—such as 10W30—that compensates for a cold or hot engine.

⊏▷ Tire Type and Size

The numbers and letters associated with tire sizes and types are complicated and confusing. Probably the most useful method of specifying tire sizes is with the "Metric P" system of numbering. The following explains the system:

P185/75R-14

- **P:**   the tire is for a passenger car

- **185:**   the width of the tire in millimeters

- **75:**   the aspect ratio—the height of the tire from the rim to the road is 75% of the width

- **R:**   it is a radial tire.

- **14:**   the wheel diameter is 14 inches (13 and 15 inches are common sizes as well)

⊏▷ Tire Pressure

In B.3, page 299, tire-pressure units are used in an example illustrating how to use the table of conversion factors. It shows that metric pressure units are related to the units commonly used in the United States by this relationship: one pound per square inch equals approximately seven kilopascals.

This means, for example, that a typical tire pressure of 26 pounds per square inch converts to approximately $7 \times 26 = 182$ kilopascals. Of course, different sized tires means different tire pressures: Consult your owner's manual or look on the tires themselves for recommended tire pressure. Then you can use your tire gauge to determine how much air is actually in your tire—and fill it up accordingly.

☞ Crash Test Index

Cars are now routinely tested for crash resistance by driving test models into fixed barriers at 35 miles an hour. Typical index numbers lie in the range 1,200 to 4,500, where lower numbers indicate less damage to the vehicle and its occupants. These index numbers can be found for all common makes of cars in various automobile or consumer publications, particularly in reports on new models of cars. The Highway Loss Data Institute also distributes them.

9

Popular Science Calculations

Scientists and engineers are constantly applying numerical methods to solve the problems they encounter. Some applications of their methods can be understood by nonspecialists—and seem worthy of inclusion here.

9.1—Satellite Rotation Periods

➩ Numbers in Orbit

Two rocky spheres, the Earth and the moon—held together by a mysterious attractive force that acts between all particles of matter—circle each other, and together circle the sun in orbits that are determined by the strength of their gravitational attractions. It was Sir Isaac Newton who established that the force between two material bodies is directly proportional to the masses of the bodies and inversely proportional to the square of the distance between them.

The radius of the Earth's orbit around the sun is about 94 million miles. This distance, which is called the "astronomical unit," is often used as a unit of measurement for distances within the solar system. (Actually, the Earth's orbit is elliptical—but its departure from circularity is less than two percent.) The average velocity of the Earth in its orbit is about nineteen miles per second, and its period of rotation is what is called a year.

 EXAMPLE

The moon's distance from the Earth is about 240,000 miles, which is about 30 times the diameter of the Earth. It travels at an orbital speed of about 0.6 miles per second, and its period of rotation is 27.3 days.

➩ Kepler's Law

Early in the 17th century, German astronomer Johannes Kepler found a simple mathematical formula that describes the period of rotation for a satellite at a certain orbital distance: "The period squared equals the distance cubed." Here, the period is measured in years and the distance is measured in astronomical units.

Using this formula, you can calculate that an artificial satellite in orbit with a rotation period of one day will be at a height of about 22,370 miles above the Earth. This is the so-called "geosynchronous" orbit because the satellite stays fixed above the same point on the surface of the Earth.

Again, the formula shows that a satellite rotating at one hundred miles altitude will circle the earth in just under ninety minutes.

Sputnik Orbits

If, as was the case for the first Soviet sputnik, the orbit of a satellite is inclined at an angle to the plane of the equator, then its path over the surface of the Earth is a complicated wavy pattern. For a trajectory inclined at sixty degrees to the equator and with an orbit period of ninety minutes, the path over the Earth appears as in Figure 9.1.

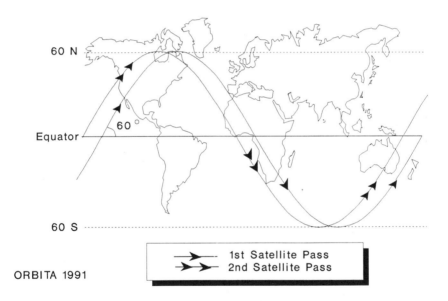

Figure 9.1—Path of a Satellite Over the Earth's Surface

The satellite crosses the equator on a northward trajectory at an angle of sixty degrees. Its path extends to sixty degrees north latitude, then re-crosses the equator and extends southward to sixty degrees south latitude. At ninety minutes, it crosses the equator again on a northward trajectory. However, during this period the Earth has rotated from west to east 22.5 degrees—so that after the second northward pass over the equator, the track is displaced by 22.5 degrees from the first pass. Consequently, the tracks over the Earth make a series of wavy lines, each one displaced westward along the equator.

In this example, the track will advance westward 22.5 degrees each ninety minutes—so that after twenty four hours and sixteen revolutions, the satellite will again pass over the same point on the Earth's surface. This means that for a photographic reconnaissance satellite viewing the Earth from an orbit at one hundred miles height, it will be necessary to have the camera scan a swath only 22.5 degrees wide to obtain full coverage once every twenty four hours in the band between sixty degrees north latitude and 60 degrees south latitude.

9.2—Distances to the Stars

✎ Astronomical Numbers

Distances between the Earth and some of the nearest objects in space can be estimated by a method involving a fixed baseline and the measurement of angles. For example, the distance from the Earth to the moon can be determined by observing a prominent feature on the moon from two points on the Earth's surface, such as the North Pole and a point on the equator. A right angle can be constructed as in Figure 9.2A at the center of the diagram of the Earth.

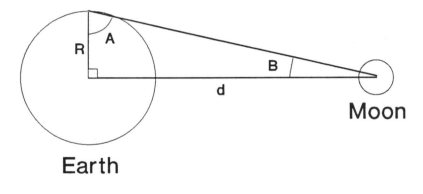

Figure 9.2A—Measuring the Distance to the Moon Using Parallax

The angle marked A in Figure 9.2A is 89.2 degrees, which means that the angle marked B is 90 – 89.2 = 0.8 degrees. Knowing the value of R, and looking up the value of the tangent of an angle of 0.8 degrees (see A.9, page 268), you can calculate the value of d, the Earth-to-moon distance. It is about 240,000 miles. The angle marked B in this case is referred to as the "parallax."

The same method can be used to measure the distance to stars, except that to make the angles big enough to be measurable, you must use a baseline longer than the Earth's radius. This is done by observing a star's position when the Earth is at A in Figure 9.2B, and another six months later when the Earth is at B. The baseline is then the radius of the Earth's orbit around the sun. In this way, the parallax of some of the nearest stars can be determined.

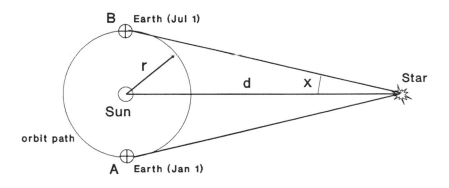

Figure 9.2B—Measuring the Distance to a Star Using Parallax

⇨ Parsecs

If the angle marked x in Figure 9.2B is one second of arc, the value of d can be found; this distance has been given the name "parsec." It is equal to about 200,000 astronomical units. (As defined in Section 9.1, page 211, an astronomical unit is the radius of the Earth's orbit around the sun, which is about 94 million miles.)

To appreciate the extreme accuracy required in measuring stellar distances using parallax, note that even for the nearest star, the angle to be measured is smaller than the angle made at the eye by a dime seen from more than a mile away. Even so, by making careful measurements and correcting for certain distorting factors, it is possible to get acceptable results for stars at less than about 50 parsecs. Beyond this distance, other, indirect methods must be used.

The nearest star is at a distance of 1.3 parsecs. Other stars are found at distances of hundreds and thousands of parsecs.

☞ Light Years

It is sometimes more convenient to use light years as units, where one light year is the distance traveled by a ray of light in one year. On this scale, the distance to the nearest star is 4.2 light years. Conversions from parsecs to light years can be made using this relationship: 1 parsec = 3.26 light years.

9.3—The Difference Between Mass and Weight

⊨ Numbers by the Pound

The concept of mass is frequently misunderstood and misused. Mass may be defined as the quantity of matter in a body, but it is more usefully defined in terms of force and acceleration. If a force is applied to any body that is free to move, it has been found experimentally that the body's speed changes in proportion to the strength of the force applied. The constant of proportionality—that is, the ratio of force to acceleration—may be defined as the mass of the body. The relationship is usually written as an equation: Force = Mass × Acceleration.

This equation applies in the general case; in a particular case of a body near the surface of the Earth, the force of gravity tends to cause a downward acceleration—and in these circumstances, the weight of the body is the mass multiplied by the acceleration due to gravity.

Much of the confusion concerning mass has arisen because it is measured in pounds, just as weight is. In fact, pounds (mass) and pounds (force) are *not* the same.

The weight of a body is a force that depends on its position relative to the Earth. If the same body is near the moon, its "weight" will be much less, although its mass remains the same. In certain orbital conditions, the weight of a body can be zero, while in high-G maneuvers in an aircraft it can be many times the force of gravity.

9.4—Einstein's Equation: $E = MC^2$

Converting Mass to Energy

Albert Einstein's famous expression for the fundamental relationship between energy and mass is probably more widely known than any other mathematical formula. It provides the mathematical basis for both the atom bomb and nuclear-power reactors.

The formula illustrates that if some means is found for converting mass into energy, there is an enormous multiplication factor (C^2)—the square of the velocity of light.

EXAMPLE **If one pound of matter could all be converted into energy, the resultant energy would be about 11 billion kilowatt hours, which would operate 30,000 color TV sets four hours a day for more than 1,000 years.**

9.5—Time and Calendars

▷ All Our Days Are Numbered

One of the classic challenges of early science was to devise a calendar that would remain accurate over many years—as a guide for sowing and harvesting crops, and for many other purposes.

Each year—the time it takes the Earth to circle the sun—is approximately 365¼ days long. By decree of Julius Caesar, the 365-day calendar, or Julian calendar, gained an extra day every fourth year to accommodate four quarter-days. These 366-day years, called leap years, were made to fall on years divisible by four (for example, 1984 and 1988). By the sixteenth century, it was evident that a further correction was needed—because the precise length of the year is eleven minutes and fourteen seconds less than 365¼ days.

Mathematicians calculated that the correction amounted to three days in 400 years—and saw that by dropping three leap years in a cycle of 400 years, all would be well.

The three leap years to delete were conveniently found by considering the century years (years ending in two zeroes) and picking as leap years only those divisible by 400. So 1600 was to be a leap year, but 1700, 1800, and 1900 would be dropped as leap years even though they would be divisible by four.

▷ Gregorian Calendar

The changeover to a calendar based on these rules began in 1582. It was called the Gregorian calendar—in honor of Pope Gregory XIII. England adopted it in 1752, when eleven days were dropped from the Julian calendar to make it conform. Greece, in 1923, was the last modern nation to make the change.

▷ There Are Fourteen Types of Calendars

Because there are two sets of calendars, for leap years and non-leap years, and seven possible calendars in each set to cover the cases of January 1 falling on a Sunday, a Monday, and so on through the week, it follows that the calendar for any particular year will be one of the fourteen possible calendars shown in Table 9.5A.

Table 9.5B is a table showing which calendar type is appropriate for any specific year.

 For the year 1992, Table 9.5B shows that Calendar Type 11 applies, and by referring to Table 9.5A, you see that 1992 is a leap year in which January 1 falls on a Wednesday.

In Table 9.5B, you can see that the pattern repeats every 400 years. For example, Calendar Type 2 applies to year 1601 and also to year 2001. Because of this, the table can be extended into the 2100s if you look under the 1700s; for the 2200s, you should look under the 1800s; and so on. For example: The calendar for year 2313 is found to be Type 1 by looking up the type for 1913. So, these tables provide a perpetual calendar for all future years.

TABLE 9.5A

Calendar Types

CALENDAR TYPE 1 (NON LEAP YEAR)

JANUARY	FEBRUARY	MARCH
S M T W T F S	S M T W T F S	S M T W T F S
1 2 3 4 5 6 7	1 2 3 4	1 2 3 4
8 9 10 11 12 13 14	5 6 7 8 9 10 11	5 6 7 8 9 10 11
15 16 17 18 19 20 21	12 13 14 15 16 17 18	12 13 14 15 16 17 18
22 23 24 25 26 27 28	19 20 21 22 23 24 25	19 20 21 22 23 24 25
29 30 31	26 27 28	26 27 28 29 30 31

APRIL	MAY	JUNE
S M T W T F S	S M T W T F S	S M T W T F S
1	1 2 3 4 5 6	1 2 3
2 3 4 5 6 7 8	7 8 9 10 11 12 13	4 5 6 7 8 9 10
9 10 11 12 13 14 15	14 15 16 17 18 19 20	11 12 13 14 15 16 17
16 17 18 19 20 21 22	21 22 23 24 25 26 27	18 19 20 21 22 23 24
23 24 25 26 27 28 29	28 29 30 31	25 26 27 28 29 30
30		

JULY	AUGUST	SEPTEMBER
S M T W T F S	S M T W T F S	S M T W T F S
1	1 2 3 4 5	1 2
2 3 4 5 6 7 8	6 7 8 9 10 11 12	3 4 5 6 7 8 9
9 10 11 12 13 14 15	13 14 15 16 17 18 19	10 11 12 13 14 15 16
16 17 18 19 20 21 22	20 21 22 23 24 25 26	17 18 19 20 21 22 23
23 24 25 26 27 28 29	27 28 29 30 31	24 25 26 27 28 29 30
30 31		

OCTOBER	NOVEMBER	DECEMBER
S M T W T F S	S M T W T F S	S M T W T F S
1 2 3 4 5 6 7	1 2 3 4	1 2
8 9 10 11 12 13 14	5 6 7 8 9 10 11	3 4 5 6 7 8 9
15 16 17 18 19 20 21	12 13 14 15 16 17 18	10 11 12 13 14 15 16
22 23 24 25 26 27 28	19 20 21 22 23 24 25	17 18 19 20 21 22 23
29 30 31	26 27 28 29 30	24 25 26 27 28 29 30
		31

CALENDAR TYPE 2 (NON LEAP YEAR)

JANUARY
```
S  M  T  W  T  F  S
      1  2  3  4  5  6
 7  8  9 10 11 12 13
14 15 16 17 18 19 20
21 22 23 24 25 26 27
28 29 30 31
```

FEBRUARY
```
S  M  T  W  T  F  S
            1  2  3
 4  5  6  7  8  9 10
11 12 13 14 15 16 17
18 19 20 21 22 23 24
25 26 27 28
```

MARCH
```
S  M  T  W  T  F  S
            1  2  3
 4  5  6  7  8  9 10
11 12 13 14 15 16 17
18 19 20 21 22 23 24
25 26 27 28 29 30 31
```

APRIL
```
S  M  T  W  T  F  S
 1  2  3  4  5  6  7
 8  9 10 11 12 13 14
15 16 17 18 19 20 21
22 23 24 25 26 27 28
29 30
```

MAY
```
S  M  T  W  T  F  S
    1  2  3  4  5
 6  7  8  9 10 11 12
13 14 15 16 17 18 19
20 21 22 23 24 25 26
27 28 29 30 31
```

JUNE
```
S  M  T  W  T  F  S
                1  2
 3  4  5  6  7  8  9
10 11 12 13 14 15 16
17 18 19 20 21 22 23
24 25 26 27 28 29 30
```

JULY
```
S  M  T  W  T  F  S
 1  2  3  4  5  6  7
 8  9 10 11 12 13 14
15 16 17 18 19 20 21
22 23 24 25 26 27 28
29 30 31
```

AUGUST
```
S  M  T  W  T  F  S
          1  2  3  4
 5  6  7  8  9 10 11
12 13 14 15 16 17 18
19 20 21 22 23 24 25
26 27 28 29 30 31
```

SEPTEMBER
```
S  M  T  W  T  F  S
                   1
 2  3  4  5  6  7  8
 9 10 11 12 13 14 15
16 17 18 19 20 21 22
23 24 25 26 27 28 29
30
```

OCTOBER
```
S  M  T  W  T  F  S
      1  2  3  4  5  6
 7  8  9 10 11 12 13
14 15 16 17 18 19 20
21 22 23 24 25 26 27
28 29 30 31
```

NOVEMBER
```
S  M  T  W  T  F  S
            1  2  3
 4  5  6  7  8  9 10
11 12 13 14 15 16 17
18 19 20 21 22 23 24
25 26 27 28 29 30
```

DECEMBER
```
S  M  T  W  T  F  S
                   1
 2  3  4  5  6  7  8
 9 10 11 12 13 14 15
16 17 18 19 20 21 22
23 24 25 26 27 28 29
30 31
```

CALENDAR TYPE 3 (NON LEAP YEAR)

	JANUARY					
S	M	T	W	T	F	S
		1	2	3	4	5
6	7	8	9	10	11	12
13	14	15	16	17	18	19
20	21	22	23	24	25	26
27	28	29	30	31		

	FEBRUARY					
S	M	T	W	T	F	S
					1	2
3	4	5	6	7	8	9
10	11	12	13	14	15	16
18	19	20	21	22	23	24
25	26	27	28			

	MARCH					
S	M	T	W	T	F	S
					1	2
3	4	5	6	7	8	9
10	11	12	13	14	15	16
17	18	19	20	21	22	23
24	25	26	27	28	29	30
31						

	APRIL					
S	M	T	W	T	F	S
	1	2	3	4	5	6
7	8	9	10	11	12	13
14	15	16	17	18	19	20
21	22	23	24	25	26	27
28	29	30				

	MAY					
S	M	T	W	T	F	S
			1	2	3	4
5	6	7	8	9	10	11
12	13	14	15	16	17	18
19	20	21	22	23	24	25
26	27	28	29	30	31	

	JUNE					
S	M	T	W	T	F	S
						1
2	3	4	5	6	7	8
9	10	11	12	13	14	15
16	17	18	19	20	21	22
23	24	25	26	27	28	29
30						

	JULY					
S	M	T	W	T	F	S
	1	2	3	4	5	6
7	8	9	10	11	12	13
14	15	16	17	18	19	20
21	22	23	24	25	26	27
28	29	30	31			

	AUGUST					
S	M	T	W	T	F	S
				1	2	3
4	5	6	7	8	9	10
11	12	13	14	15	16	17
18	19	20	21	22	23	24
25	26	27	28	29	30	31

	SEPTEMBER					
S	M	T	W	T	F	S
1	2	3	4	5	6	7
8	9	10	11	12	13	14
15	16	17	18	19	20	21
22	23	24	25	26	27	28
29	30					

	OCTOBER					
S	M	T	W	T	F	S
		1	2	3	4	5
6	7	8	9	10	11	12
13	14	15	16	17	18	19
20	21	22	23	24	25	26
27	28	29	30	31		

	NOVEMBER					
S	M	T	W	T	F	S
					1	2
3	4	5	6	7	8	9
9	10	11	12	13	14	15
16	17	18	19	20	21	22
23	24	25	26	27	28	29
30	31					

	DECEMBER					
S	M	T	W	T	F	S
1	2	3	4	5	6	7
8	9	10	11	12	13	14
15	16	17	18	19	20	21
22	23	24	25	26	27	28
29	30	31				

CALENDAR TYPE 4 (NON LEAP YEAR)

JANUARY
```
 S  M  T  W  T  F  S
          1  2  3  4
 5  6  7  8  9 10 11
12 13 14 15 16 17 18
19 20 21 22 23 24 25
26 27 28 29 30 31
```

FEBRUARY
```
 S  M  T  W  T  F  S
                   1
 2  3  4  5  6  7  8
 9 10 11 12 13 14 15
16 17 18 19 20 21 22
23 24 25 26 27 28
```

MARCH
```
 S  M  T  W  T  F  S
                   1
 2  3  4  5  6  7  8
 9 10 11 12 13 14 15
15 16 17 18 19 20 21
22 23 24 25 26 27 28
29 30 31
```

APRIL
```
 S  M  T  W  T  F  S
       1  2  3  4  5
 6  7  8  9 10 11 12
13 14 15 16 17 18 19
20 21 22 23 24 25 26
27 28 29 30
```

MAY
```
 S  M  T  W  T  F  S
                1  2  3
 4  5  6  7  8  9 10
11 12 13 14 15 16 17
18 19 20 21 22 23 24
25 26 27 28 29 30 31
```

JUNE
```
 S  M  T  W  T  F  S
 1  2  3  4  5  6  7
 8  9 10 11 12 13 14
15 16 17 18 19 20 21
22 23 24 25 26 27 28
29 30
```

JULY
```
 S  M  T  W  T  F  S
       1  2  3  4  5
 6  7  8  9 10 11 12
13 14 15 16 17 18 19
20 21 22 23 24 25 26
27 28 29 30 31
```

AUGUST
```
 S  M  T  W  T  F  S
                1  2
 3  4  5  6  7  8  9
10 11 12 13 14 15 16
17 18 19 20 21 22 23
24 25 26 27 28 29 30
31
```

SEPTEMBER
```
 S  M  T  W  T  F  S
    1  2  3  4  5  6
 7  8  9 10 11 12 13
14 15 16 17 18 19 20
21 22 23 24 25 26 27
28 29 30
```

OCTOBER
```
 S  M  T  W  T  F  S
          1  2  3  4
 5  6  7  8  9 10 11
12 13 14 15 16 17 18
19 20 21 22 23 24 25
26 27 28 29 30 31
```

NOVEMBER
```
 S  M  T  W  T  F  S
                   1
 2  3  4  5  6  7  8
 9 10 11 12 13 14 15
16 17 18 19 20 21 22
23 24 25 26 27 28 29
30
```

DECEMBER
```
 S  M  T  W  T  F  S
    1  2  3  4  5  6
 7  8  9 10 11 12 13
14 15 16 17 18 19 20
21 22 23 24 25 26 27
28 29 30 31
```

CALENDAR TYPE 5 (NON LEAP YEAR)

JANUARY
S	M	T	W	T	F	S
					1	2	3
4	5	6	7	8	9	10
11	12	13	14	15	16	17
18	19	20	21	22	23	24
25	26	27	28	29	30	31

FEBRUARY
S	M	T	W	T	F	S
1	2	3	4	5	6	7
8	9	10	11	12	13	14
15	16	17	18	19	20	21
22	23	24	25	26	27	28

MARCH
S	M	T	W	T	F	S
1	2	3	4	5	6	7
8	9	10	11	12	13	14
15	16	17	18	19	20	21
22	23	24	25	26	27	28
29	30	31				

APRIL
S	M	T	W	T	F	S
				1	2	3	4
5	6	7	8	9	10	11
12	13	14	15	16	17	18
19	20	21	22	23	24	25
26	27	28	29	30		

MAY
S	M	T	W	T	F	S
					1	2
3	4	5	6	7	8	9
10	11	12	13	14	15	16
17	18	19	20	21	22	23
24	25	26	27	28	29	30
31						

JUNE
S	M	T	W	T	F	S
	1	2	3	4	5	6
7	8	9	10	11	12	13
14	15	16	17	18	19	20
21	22	23	24	25	26	27
28	29	30				

JULY
S	M	T	W	T	F	S
				1	2	3	4
5	6	7	8	9	10	11
12	13	14	15	16	17	18
19	20	21	22	23	24	25
26	27	28	29	29	30	31

AUGUST
S	M	T	W	T	F	S
						1
2	3	4	5	6	7	8
9	10	11	12	13	14	15
16	17	18	19	20	21	22
23	24	25	26	27	28	29
30	31					

SEPTEMBER
S	M	T	W	T	F	S
		1	2	3	4	5
6	7	8	9	10	11	12
13	14	15	16	17	18	19
20	21	22	23	24	25	26
27	28	29	30			

OCTOBER
S	M	T	W	T	F	S
					1	2	3
4	5	6	7	8	9	10
11	12	13	14	15	16	17
18	19	20	21	22	23	24
25	26	27	28	29	30	31

NOVEMBER
S	M	T	W	T	F	S
1	2	3	4	5	6	7
8	9	10	11	12	13	14
15	16	17	18	19	20	21
22	23	24	25	26	27	28
29	30					

DECEMBER
S	M	T	W	T	F	S
		1	2	3	4	5
6	7	8	9	10	11	12
13	14	15	16	17	18	19
20	21	22	23	24	25	26
27	28	29	30	31		

CALENDAR TYPE 6 (NON LEAP YEAR)

JANUARY
S	M	T	W	T	F	S
					1	2
3	4	5	6	7	8	9
10	11	12	13	14	15	16
17	18	19	20	21	22	23
24	25	26	27	28	29	30
31						

FEBRUARY
S	M	T	W	T	F	S
1	2	3	4	5	6	
7	8	9	10	11	12	13
14	15	16	17	18	19	20
21	22	23	24	25	26	27
28						

MARCH
S	M	T	W	T	F	S
1	2	3	4	5	6	
7	8	9	10	11	12	13
14	15	16	17	18	19	20
21	22	23	24	25	26	27
28	29	30	31			

APRIL
S	M	T	W	T	F	S
				1	2	3
4	5	6	7	8	9	10
11	12	13	14	15	16	17
18	19	20	21	22	23	24
25	26	27	28	29	30	

MAY
S	M	T	W	T	F	S
						1
2	3	4	5	6	7	8
9	10	11	12	13	14	15
16	17	18	19	20	21	22
23	24	25	26	27	28	29
30	31					

JUNE
S	M	T	W	T	F	S
	1	2	3	4	5	
6	7	8	9	10	11	12
13	14	15	16	17	18	19
20	21	22	23	24	25	26
27	28	29	30			

JULY
S	M	T	W	T	F	S
				1	2	3
4	5	6	7	8	9	10
11	12	13	14	15	16	17
18	19	20	21	22	23	24
25	26	27	28	29	30	31

AUGUST
S	M	T	W	T	F	S
1	2	3	4	5	6	7
8	9	10	11	12	13	14
15	16	17	18	19	20	21
22	23	24	25	26	27	28
29	30	31				

SEPTEMBER
S	M	T	W	T	F	S
			1	2	3	4
5	6	7	8	9	10	11
12	13	14	15	16	17	18
19	20	21	22	23	24	25
26	27	28	29	30		

OCTOBER
S	M	T	W	T	F	S
					1	2
3	4	5	6	7	8	9
10	11	12	13	14	15	16
17	18	19	20	21	22	23
24	25	26	27	28	29	30
31						

NOVEMBER
S	M	T	W	T	F	S
1	2	3	4	5	6	
7	8	9	10	11	12	13
14	15	16	17	18	19	20
21	22	23	24	25	26	27
28	29	30				

DECEMBER
S	M	T	W	T	F	S
			1	2	3	4
5	6	7	8	9	10	11
12	13	14	15	16	17	18
19	20	21	22	23	24	25
26	27	28	29	30	31	

CALENDAR TYPE 7 (NON LEAP YEAR)

	JANUARY					
S	M	T	W	T	F	S
						1
2	3	4	5	6	7	8
9	10	11	12	13	14	15
16	17	18	19	20	21	22
23	24	25	26	27	28	29
30	31					

	FEBRUARY					
S	M	T	W	T	F	S
		1	2	3	4	5
6	7	8	9	10	11	12
13	14	15	16	17	18	19
20	21	22	23	24	25	26
27	28					

	MARCH					
S	M	T	W	T	F	S
		1	2	3	4	5
6	7	8	9	10	11	12
13	14	15	16	17	18	19
20	21	22	23	24	25	26
27	28	29	30	31		

	APRIL					
S	M	T	W	T	F	S
					1	2
3	4	5	6	7	8	9
10	11	12	13	14	15	16
17	18	19	20	21	22	23
24	25	26	27	28	29	30

	MAY					
S	M	T	W	T	F	S
1	2	3	4	5	6	7
8	9	10	11	12	13	14
15	16	17	18	19	20	21
22	23	24	25	26	27	28
29	30	31				

	JUNE					
S	M	T	W	T	F	S
			1	2	3	4
5	6	7	8	9	10	11
12	13	14	15	16	17	18
19	20	21	22	23	24	25
26	27	28	29	30		

	JULY					
S	M	T	W	T	F	S
					1	2
3	4	5	6	7	8	9
10	11	12	13	14	15	16
17	18	19	20	21	22	23
24	25	26	27	28	29	30
31						

	AUGUST					
S	M	T	W	T	F	S
	1	2	3	4	5	6
7	8	9	10	11	12	13
14	15	16	17	18	19	20
21	22	23	24	25	26	27
28	29	30	31			

	SEPTEMBER					
S	M	T	W	T	F	S
				1	2	3
4	5	6	7	8	9	10
11	12	13	14	15	16	17
18	19	20	21	22	23	24
25	26	27	28	29	30	

	OCTOBER					
S	M	T	W	T	F	S
						1
2	3	4	5	6	7	8
9	10	11	12	13	14	15
16	17	18	19	20	21	22
23	24	25	26	27	28	29
30	31					

	NOVEMBER					
S	M	T	W	T	F	S
		1	2	3	4	5
6	7	8	9	10	11	12
13	14	15	16	17	18	19
20	21	22	23	24	25	26
27	28	29	30			

	DECEMBER					
S	M	T	W	T	F	S
				1	2	3
4	5	6	7	8	9	10
11	12	13	14	15	16	17
18	19	20	21	22	23	24
25	26	27	28	29	30	31

CALENDAR TYPE 8 (LEAP YEAR)

JANUARY
S	M	T	W	T	F	S
1	2	3	4	5	6	7
8	9	10	11	12	13	14
15	16	17	18	19	20	21
22	23	24	25	26	27	28
29	30	31				

FEBRUARY
S	M	T	W	T	F	S
				1	2	3	4
5	6	7	8	9	10	11
12	13	14	15	16	17	18
19	20	21	22	23	24	25
26	27	28	29			

MARCH
S	M	T	W	T	F	S
					1	2	3
4	5	6	7	8	9	10
11	12	13	14	15	16	17
18	19	20	21	22	23	24
25	26	27	28	29	30	31

APRIL
S	M	T	W	T	F	S
1	2	3	4	5	6	7
8	9	10	11	12	13	14
15	16	17	18	19	20	21
22	23	24	25	26	27	28
29	30					

MAY
S	M	T	W	T	F	S
	1	2	3	4	5	
6	7	8	9	10	11	12
13	14	15	16	17	18	19
20	21	22	23	24	25	26
27	28	29	30	31		

JUNE
S	M	T	W	T	F	S
						1
3	4	5	6	7	8	9
10	11	12	13	14	15	16
17	18	19	20	21	22	23
24	25	26	27	28	29	30

JULY
S	M	T	W	T	F	S
1	2	3	4	5	6	7
8	9	10	11	12	13	14
15	16	17	18	19	20	21
22	23	24	25	26	27	28
29	30	31				

AUGUST
S	M	T	W	T	F	S
				1	2	3	4
5	6	7	8	9	10	11
12	13	14	15	16	17	18
19	20	21	22	23	24	25
26	27	28	29	30	31	

SEPTEMBER
S	M	T	W	T	F	S
						1
2	3	4	5	6	7	8
9	10	11	12	13	14	15
16	17	18	19	20	21	22
23	24	25	26	27	28	29
30						

OCTOBER
S	M	T	W	T	F	S
	1	2	3	4	5	6
7	8	9	10	11	12	13
14	15	16	17	18	19	20
21	22	23	24	25	26	27
28	29	30	31			

NOVEMBER
S	M	T	W	T	F	S
				1	2	3
4	5	6	7	8	9	10
11	12	13	14	15	16	17
18	19	20	21	22	23	24
25	26	27	28	29	30	

DECEMBER
S	M	T	W	T	F	S
						1
2	3	4	5	6	7	8
9	10	11	12	13	14	15
16	17	18	19	20	21	22
23	24	25	26	27	28	29
30	31					

CALENDAR TYPE 9 (LEAP YEAR)

JANUARY
```
S  M  T  W  T  F  S
      1  2  3  4  5  6
7  8  9 10 11 12 13
14 15 16 17 18 19 20
21 22 23 24 25 26 27
28 29 30 31
```

FEBRUARY
```
S  M  T  W  T  F  S
               1  2  3
4  5  6  7  8  9 10
11 12 13 14 15 16 17
18 19 20 21 22 23 24
25 26 27 28 29
```

MARCH
```
S  M  T  W  T  F  S
                  1  2
3  4  5  6  7  8  9
10 11 12 13 14 15 16
17 18 19 20 21 22 23
24 25 26 27 28 29 30
31
```

APRIL
```
S  M  T  W  T  F  S
      1  2  3  4  5  6
7  8  9 10 11 12 13
14 15 16 17 18 19 20
21 22 23 24 25 26 27
28 29 30
```

MAY
```
S  M  T  W  T  F  S
            1  2  3  4
5  6  7  8  9 10 11
12 13 14 15 16 17 18
19 20 21 22 23 24 25
26 27 28 29 30 31
```

JUNE
```
S  M  T  W  T  F  S
                     1
2  3  4  5  6  7  8
9 10 11 12 13 14 15
16 17 18 19 20 21 22
23 24 25 26 27 28 29
30
```

JULY
```
S  M  T  W  T  F  S
      1  2  3  4  5  6
7  8  9 10 11 12 13
14 15 16 17 18 19 20
21 22 23 24 25 26 27
28 29 30 31
```

AUGUST
```
S  M  T  W  T  F  S
               1  2  3
4  5  6  7  8  9 10
11 12 13 14 15 16 17
18 19 20 21 22 23 24
25 26 27 28 29 30 31
```

SEPTEMBER
```
S  M  T  W  T  F  S
1  2  3  4  5  6  7
8  9 10 11 12 13 14
15 16 17 18 19 20 21
22 23 24 25 26 27 28
29 30
```

OCTOBER
```
S  M  T  W  T  F  S
         1  2  3  4  5
6  7  8  9 10 11 12
13 14 15 16 17 18 19
20 21 22 23 24 25 26
27 28 29 30 31
```

NOVEMBER
```
S  M  T  W  T  F  S
                  1  2
3  4  5  6  7  8  9
9 10 11 12 13 14 15
16 17 18 19 20 21 22
23 24 25 26 27 28 29
30 31
```

DECEMBER
```
S  M  T  W  T  F  S
1  2  3  4  5  6  7
8  9 10 11 12 13 14
15 16 17 18 19 20 21
22 23 24 25 26 27 28
29 30 31
```

CALENDAR TYPE 10 (LEAP YEAR)

JANUARY
S	M	T	W	T	F	S
		1	2	3	4	5
6	7	8	9	10	11	12
13	14	15	16	17	18	19
20	21	22	23	24	25	26
27	28	29	30	31		

FEBRUARY
S	M	T	W	T	F	S
					1	2
3	4	5	6	7	8	9
10	11	12	13	14	15	16
17	18	19	20	21	22	23
24	25	26	27	28	29	

MARCH
S	M	T	W	T	F	S
						1
2	3	4	5	6	7	8
9	10	11	12	13	14	15
16	17	18	19	20	21	22
23	24	25	26	27	28	29
30	31					

APRIL
S	M	T	W	T	F	S
		1	2	3	4	5
6	7	8	9	10	11	12
13	14	15	16	17	18	19
20	21	22	23	24	25	26
27	28	29	30			

MAY
S	M	T	W	T	F	S
				1	2	3
4	5	6	7	8	9	10
11	12	13	14	15	16	17
18	19	20	21	22	23	24
25	26	27	28	29	30	31

JUNE
S	M	T	W	T	F	S
1	2	3	4	5	6	7
8	9	10	11	12	13	14
15	16	17	18	19	20	21
22	23	24	25	26	27	28
29	30					

JULY
S	M	T	W	T	F	S
		1	2	3	4	5
6	7	8	9	10	11	12
13	14	15	16	17	18	19
20	21	22	23	24	25	26
27	28	29	30	31		

AUGUST
S	M	T	W	T	F	S
					1	2
3	4	5	6	7	8	9
10	11	12	13	14	15	16
17	18	19	20	21	22	23
24	25	26	27	28	29	30
31						

SEPTEMBER
S	M	T	W	T	F	S
	1	2	3	4	5	6
7	8	9	10	11	12	13
14	15	16	17	18	19	20
21	22	23	24	25	26	27
28	29	30				

OCTOBER
S	M	T	W	T	F	S
			1	2	3	4
5	6	7	8	9	10	11
12	13	14	15	16	17	18
19	20	21	22	23	24	25
26	27	28	29	30	31	

NOVEMBER
S	M	T	W	T	F	S
						1
2	3	4	5	6	7	8
9	10	11	12	13	14	15
16	17	18	19	20	21	22
23	24	25	26	27	28	29
30						

DECEMBER
S	M	T	W	T	F	S
	1	2	3	4	5	6
7	8	9	10	11	12	13
14	15	16	17	18	19	20
21	22	23	24	25	26	27
28	29	30	31			

CALENDAR TYPE 11 (LEAP YEAR)

JANUARY
```
S  M  T  W  T  F  S
         1  2  3  4
 5  6  7  8  9 10 11
12 13 14 15 16 17 18
19 20 21 22 23 24 25
26 27 28 29 30 31
```

FEBRUARY
```
S  M  T  W  T  F  S
                  1
 2  3  4  5  6  7  8
 9 10 11 12 13 14 15
16 17 18 19 20 21 22
23 24 25 26 27 28 29
```

MARCH
```
S  M  T  W  T  F  S
 1  2  3  4  5  6  7
 8  9 10 11 12 13 14
15 16 17 18 19 20 21
22 23 24 25 26 27 28
29 30 31
```

APRIL
```
S  M  T  W  T  F  S
         1  2  3  4
 5  6  7  8  9 10 11
12 13 14 15 16 17 18
19 20 21 22 23 24 25
26 27 28 29 30
```

MAY
```
S  M  T  W  T  F  S
               1  2
 3  4  5  6  7  8  9
10 11 12 13 14 15 16
17 18 19 20 21 22 23
24 25 26 27 28 29 30
31
```

JUNE
```
S  M  T  W  T  F  S
 1  2  3  4  5  6
 7  8  9 10 11 12 13
14 15 16 17 18 19 20
21 22 23 24 25 26 27
28 29 30
```

JULY
```
S  M  T  W  T  F  S
         1  2  3  4
 5  6  7  8  9 10 11
12 13 14 15 16 17 18
19 20 21 22 23 24 25
26 27 28 29 30 31
```

AUGUST
```
S  M  T  W  T  F  S
                  1
 2  3  4  5  6  7  8
 9 10 11 12 13 14 15
16 17 18 19 20 21 22
23 24 25 26 27 28 29
30 31
```

SEPTEMBER
```
S  M  T  W  T  F  S
    1  2  3  4  5
 6  7  8  9 10 11 12
13 14 15 16 17 18 19
20 21 22 23 24 25 26
27 28 29 30
```

OCTOBER
```
S  M  T  W  T  F  S
            1  2  3
 4  5  6  7  8  9 10
11 12 13 14 15 16 17
18 19 20 21 22 23 24
25 26 27 28 29 30 31
```

NOVEMBER
```
S  M  T  W  T  F  S
 1  2  3  4  5  6  7
 8  9 10 11 12 13 14
15 16 17 18 19 20 21
22 23 24 25 26 27 28
29 30
```

DECEMBER
```
S  M  T  W  T  F  S
       1  2  3  4  5
 6  7  8  9 10 11 12
13 14 15 16 17 18 19
20 21 22 23 24 25 26
27 28 29 30 31
```

CALENDAR TYPE 12 (LEAP YEAR)

JANUARY
S	M	T	W	T	F	S
				1	2	3
4	5	6	7	8	9	10
11	12	13	14	15	16	17
18	19	20	21	22	23	24
25	26	27	28	29	30	31

FEBRUARY
S	M	T	W	T	F	S
1	2	3	4	5	6	7
8	9	10	11	12	13	14
15	16	17	18	19	20	21
22	23	24	25	26	27	28
29						

MARCH
S	M	T	W	T	F	S
	1	2	3	4	5	6
7	8	9	10	11	12	13
14	15	16	17	18	19	20
21	22	23	24	25	26	27
28	29	30	31			

APRIL
S	M	T	W	T	F	S
				1	2	3
4	5	6	7	8	9	10
11	12	13	14	15	16	17
18	19	20	21	22	23	24
25	26	27	28	29	30	

MAY
S	M	T	W	T	F	S
						1
2	3	4	5	6	7	8
9	10	11	12	13	14	15
16	17	18	19	20	21	22
23	24	25	26	27	28	29
30	31					

JUNE
S	M	T	W	T	F	S
		1	2	3	4	5
6	7	8	9	10	11	12
13	14	15	16	17	18	19
20	21	22	23	24	25	26
27	28	29	30			

JULY
S	M	T	W	T	F	S
				1	2	3
4	5	6	7	8	9	10
11	12	13	14	15	16	17
18	19	20	21	22	23	24
25	26	27	28	29	30	31

AUGUST
S	M	T	W	T	F	S
1	2	3	4	5	6	7
8	9	10	11	12	13	14
15	16	17	18	19	20	21
22	23	24	25	26	27	28
29	30	31				

SEPTEMBER
S	M	T	W	T	F	S
			1	2	3	4
5	6	7	8	9	10	11
12	13	14	15	16	17	18
19	20	21	22	23	24	25
26	27	28	29	30		

OCTOBER
S	M	T	W	T	F	S
					1	2
3	4	5	6	7	8	9
10	11	12	13	14	15	16
17	18	19	20	21	22	23
24	25	26	27	28	29	30
31						

NOVEMBER
S	M	T	W	T	F	S
	1	2	3	4	5	6
7	8	9	10	11	12	13
14	15	16	17	18	19	20
21	22	23	24	25	26	27
28	29	30				

DECEMBER
S	M	T	W	T	F	S
		1	2	3	4
5	6	7	8	9	10	11
12	13	14	15	16	17	18
19	20	21	22	23	24	25
26	27	28	29	30	31	

CALENDAR TYPE 13 (LEAP YEAR)

JANUARY
S	M	T	W	T	F	S
					1	2
3	4	5	6	7	8	9
10	11	12	13	14	15	16
17	18	19	20	21	22	23
24	25	26	27	28	29	30
31						

FEBRUARY
S	M	T	W	T	F	S
1	2	3	4	5	6	
7	8	9	10	11	12	13
14	15	16	17	18	19	20
21	22	23	24	25	26	27
28	29					

MARCH
S	M	T	W	T	F	S
		1	2	3	4	5
6	7	8	9	10	11	12
13	14	15	16	17	18	19
20	21	22	23	24	25	26
27	28	29	30	31		

APRIL
S	M	T	W	T	F	S
					1	2
3	4	5	6	7	8	9
10	11	12	13	14	15	16
17	18	19	20	21	22	23
24	25	26	27	28	29	30

MAY
S	M	T	W	T	F	S
1	2	3	4	5	6	7
8	9	10	11	12	13	14
15	16	17	18	19	20	21
22	23	24	25	26	27	28
29	30	31				

JUNE
S	M	T	W	T	F	S
			1	2	3	4
5	6	7	8	9	10	11
12	13	14	15	16	17	18
19	20	21	22	23	24	25
26	27	28	29	30		

JULY
S	M	T	W	T	F	S
					1	2
3	4	5	6	7	8	9
10	11	12	13	14	15	16
17	18	19	20	21	22	23
24	25	26	27	28	29	30
31						

AUGUST
S	M	T	W	T	F	S
1	2	3	4	5	6	
7	8	9	10	11	12	13
14	15	16	17	18	19	20
21	22	23	24	25	26	27
28	29	30	31			

SEPTEMBER
S	M	T	W	T	F	S
				1	2	3
4	5	6	7	8	9	10
11	12	13	14	15	16	17
18	19	20	21	22	23	24
25	26	27	28	29	30	

OCTOBER
S	M	T	W	T	F	S
						1
2	3	4	5	6	7	8
9	10	11	12	13	14	15
16	17	18	19	20	21	22
23	24	25	26	27	28	29
30	31					

NOVEMBER
S	M	T	W	T	F	S
		1	2	3	4	5
6	7	8	9	10	11	12
13	14	15	16	17	18	19
20	21	22	23	24	25	26
27	28	29	30			

DECEMBER
S	M	T	W	T	F	S
				1	2	3
4	5	6	7	8	9	10
11	12	13	14	15	16	17
18	19	20	21	22	23	24
25	26	27	28	29	30	31

CALENDAR TYPE 14 (LEAP YEAR)

JANUARY
```
S  M  T  W  T  F  S
                  1
 2  3  4  5  6  7  8
 9 10 11 12 13 14 15
16 17 18 19 20 21 22
23 24 25 26 27 28 29
```

FEBRUARY
```
S  M  T  W  T  F  S
    1  2  3  4  5
 6  7  8  9 10 11 12
13 14 15 16 17 18 19
20 21 22 23 24 25 26
27 28 29
```

MARCH
```
S  M  T  W  T  F  S
          1  2  3  4
 5  6  7  8  9 10 11
12 13 14 15 16 17 18
19 20 21 22 23 24 25
26 27 28 29 30 31
```

APRIL
```
S  M  T  W  T  F  S
                  1
 2  3  4  5  6  7  8
 9 10 11 12 13 14 15
16 17 18 19 20 21 22
23 24 25 26 27 28 29
30
```

MAY
```
S  M  T  W  T  F  S
 1  2  3  4  5  6
 7  8  9 10 11 12 13
14 15 16 17 18 19 20
21 22 23 24 25 26 27
28 29 30 31
```

JUNE
```
S  M  T  W  T  F  S
             1  2  3
 4  5  6  7  8  9 10
11 12 13 14 15 16 17
18 19 20 21 22 23 24
25 26 27 28 29 30
```

JULY
```
S  M  T  W  T  F  S
                  1
 2  3  4  5  6  7  8
 9 10 11 12 13 14 15
16 17 18 19 20 21 22
23 24 25 26 27 28 29
30 31
```

AUGUST
```
S  M  T  W  T  F  S
    1  2  3  4  5
 6  7  8  9 10 11 12
13 14 15 16 17 18 19
20 21 22 23 24 25 26
27 28 29 30 31
```

SEPTEMBER
```
S  M  T  W  T  F  S
                1  2
 3  4  5  6  7  8  9
10 11 12 13 14 15 16
17 18 19 20 21 22 23
24 25 26 27 28 29 30
```

OCTOBER
```
S  M  T  W  T  F  S
 1  2  3  4  5  6  7
 8  9 10 11 12 13 14
15 16 17 18 19 20 21
22 23 24 25 26 27 28
29 30 31
```

NOVEMBER
```
S  M  T  W  T  F  S
       1  2  3  4
 5  6  7  8  9 10 11
12 13 14 15 16 17 18
19 20 21 22 23 24 25
26 27 28 29 30
```

DECEMBER
```
S  M  T  W  T  F  S
                1  2
 3  4  5  6  7  8  9
10 11 12 13 14 15 16
17 18 19 20 21 22 23
24 25 26 27 28 29 30
31
```

TABLE 9.5B

Calendar Types for Different Years

YEAR										TYPE
1589	1645	—	1741	1797	1837	1893	1933	1989	2045	1
1590	1646	—	1742	1798	1838	1894	1934	1990	2046	2
1591	1647	—	1743	1799	1839	1895	1935	1991	2047	3
1592	1648	—	1744	—	1840	1896	1936	1992	2048	11
1593	1649	—	1745	—	1841	1897	1937	1993	2049	6
1594	1650	—	1746	—	1842	1898	1938	1994	2050	7
1595	1651	—	1747	—	1843	1899	1939	1995	2051	1
1596	1652	—	1748	—	1844	—	1940	1996	2052	9
1597	1653	—	1749	—	1845	—	1941	1997	2053	4
1598	1654	—	1750	—	1846	—	1942	1998	2054	5
1599	1655	1700	1751	—	1847	—	1943	1999	2055	6
1600	1656	—	1752	—	1848	—	1944	2000	2056	14
1601	1657	—	1753	—	1849	—	1945	2001	2057	2
1602	1658	—	1754	—	1850	—	1946	2002	2058	3
1603	1659	—	1755	1800	1851	—	1947	2003	2059	4
1604	1660	—	1756	—	1852	—	1948	2004	2060	12
1605	1661	1701	1757	—	1853	—	1949	2005	2061	7
1606	1662	1702	1758	—	1854	—	1950	2006	2062	1
1607	1663	1703	1759	—	1855	1900	1951	2007	2063	2
1608	1664	1704	1760	—	1856	—	1952	2008	2064	10
1609	1665	1705	1761	1801	1857	—	1953	2009	2065	5
1610	1666	1706	1762	1802	1858	—	1954	2010	2066	6
1611	1667	1707	1763	1803	1859	—	1955	2011	2067	7
1612	1668	1708	1764	1804	1860	—	1956	2012	2068	8
1613	1669	1709	1765	1805	1861	1901	1957	2013	2069	3
1614	1670	1710	1766	1806	1862	1902	1958	2014	2070	4
1615	1671	1711	1767	1807	1863	1903	1959	2015	2071	5
1616	1672	1712	1768	1808	1864	1904	1960	2016	2072	13
1617	1673	1713	1769	1809	1865	1905	1961	2017	2073	1
1618	1674	1714	1770	1810	1866	1906	1962	2018	2074	2
1619	1675	1715	1771	1811	1867	1907	1963	2019	2075	3
1620	1676	1716	1772	1812	1868	1908	1964	2020	2076	11
1621	1677	1717	1773	1813	1869	1909	1965	2021	2077	6
1622	1678	1718	1774	1814	1870	1910	1966	2022	2078	7
1623	1679	1719	1775	1815	1871	1911	1967	2023	2079	1
1624	1680	1720	1776	1816	1872	1912	1968	2024	2080	9
1625	1681	1721	1777	1817	1873	1913	1969	2025	2081	4

					YEAR					TYPE
	1626	1682	1722	1778	1818	1874	1914	1970	2026	2082	5
	1627	1683	1723	1779	1819	1875	1915	1971	2027	2083	6
	1628	1684	1724	1780	1820	1876	1916	1972	2028	2084	14
	1629	1685	1725	1781	1821	1877	1917	1973	2029	2085	2
	1630	1686	1726	1782	1822	1878	1918	1974	2030	2086	3
	1631	1687	1727	1783	1823	1879	1919	1975	2031	2087	4
	1632	1688	1728	1784	1824	1880	1920	1976	2032	2088	12
	1633	1689	1729	1785	1825	1881	1921	1977	2033	2089	7
	1634	1690	1730	1786	1826	1882	1922	1978	2034	2090	1
	1635	1691	1731	1787	1827	1883	1923	1979	2035	2091	2
	1636	1692	1732	1788	1828	1884	1924	1980	2036	2092	10
—	1637	1693	1733	1789	1829	1885	1925	1981	2037	2093	5
—	1638	1694	1734	1790	1830	1886	1926	1982	2038	2094	6
1583	1639	1695	1735	1791	1831	1887	1927	1983	2039	2095	7
1584	1640	1696	1736	1792	1832	1888	1928	1984	2040	2096	8
1585	1641	1697	1737	1793	1833	1889	1929	1985	2041	2097	3
1586	1642	1698	1738	1794	1834	1890	1930	1986	2042	2098	4
1587	1643	1699	1739	1795	1835	1891	1931	1987	2043	2099	5
1588	1644	—	1740	1796	1836	1892	1932	1988	2044	—	13

9.6—Radioactive Decay

Half-dead Numbers

Some of the natural elements spontaneously emit nuclear radiation; they are said to be radioactive. Other elements, when exposed in nuclear reactors, can be made artificially radioactive. The radioactive form of cobalt, produced by exposure in a reactor, is a particularly useful substance, since the high-energy rays that it emits can be used in the treatment of cancer.

After a period that may vary from fractions of a second to many years, radioactive substances decay and form stable elements. The rate of decay of all radioactive substances follows an exponential law which says that the rate of decay depends on the amount of the substance remaining.

Half-lives

The simplest way to describe the rate of decay is in terms of the "half-life," T, which is the time required for half of the original atoms of the substance to decay. The simple formula for the process is then: $T = .693k$, where k is a constant that is characteristic of the substance.

A typical decay curve is shown in Figure 9.6.

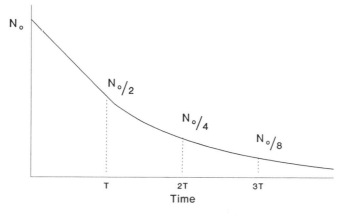

Figure 9.6—Radioactive Decay Curve

The best-known naturally radioactive substance is uranium, which decays through a series of steps, including the production of radium with a half-life of 1,620 years and radon, a gaseous form of radium, with a half-life of 3.82 days. (See Section 5.7, page 108.)

9.7—Compass Errors

⬭ Numbers for Navigators

It is common knowledge that a magnetic compass does not point directly north. If you plan to rely on it to find your way through a wilderness area, or on the water or in the air, you should find out how much it needs to be corrected for the area with which you are concerned.

Roughly, in the central United States, at about the longitude of Chicago, you don't need to make a correction. Going east, the correction increases to about ten degrees west around Washington, D.C., and fifteen degrees west around Boston. (This means, for example, that around Boston, your compass will point fifteen degrees west of true north). Going west from Chicago, the correction is ten degrees east around the longitude of Denver, fifteen degrees east around Los Angeles and twenty degrees east around Seattle. If you use a compass in Alaska, you will need to correct it by twenty-five to thirty degrees east.

If it will help you visualize the situation, you can picture a Magnetic North Pole situated in the Canadian Arctic islands (around 76N, 101W). But it is better to think of the compass as a device that is affected by local variations in the Earth's magnetic field due to many factors, including where you are, what iron ore there might be nearby, what metal there might be nearby in the engine of your car or boat or plane, or in your pocket. The point is you should be careful to check your compass for errors before you have to rely on it.

You can do this best by taking it out at night, lining up with Polaris, the North Star, and reading how much the compass error is where you are going to use it.

Appendix A

Tools You Might Need

I n Appendix A, you'll find some of the basic mathematical tools you'll need to solve the practical numerical problems you'll encounter in everyday life. The topics we cover here are supplemental to those in the main text of the book. If you want to go into a topic in more detail, we offer suggestions for further reading; at your local library, you will find a great variety of mathematical textbooks at every level of complexity.

Appendix B expands on units of measurement that come into most numerical problems, and Appendix C expands on financial mathematics.

A.1—Using Tables, Functions and Graphs

Most numerical problems can be stated in terms of equations and functions, and the results can be presented in the form of tables and graphs. Often, it is convenient to use letter symbols to represent numerical quantities when making general statements. A simple example can be used to illustrate these concepts: Suppose you want to buy a carpet to cover a square floor, and you need to calculate how many square feet or square meters you should buy. The area can be expressed by the simple equation:

$$A = s \times s \text{ or } A = s^2$$

where A represents the area and s represents the length of each side.

In this case, the area is said to be a "function" of the side length because there is a value A corresponding to each value s. Using the equation and calculating various values of A, the results can be presented either in the form of a table or a graph, as follows:

Length in Feet	Area in Sq. Feet
1	1
2	4
3	9
4	16
5	25

Table A.1—Area of a Square as a Function of Side Length

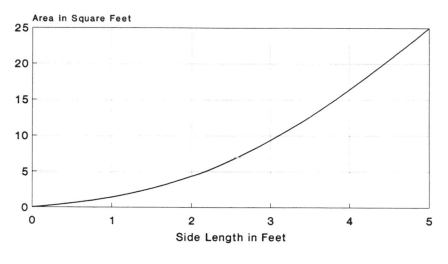

Figure A.1—Area of a Square as a Function of Side Length

Note that if you had the results of this calculation, in the form of either a table or a graph, you would not need to use the original equation, but you would be able to read the answer when you had measured the side of your room.

A.2—How to Interpolate to Find Intermediate Values

It will often be necessary to "interpolate" between values in tables. Intermediate results can be read directly from graphs, so it is not necessary to interpolate when you have a graph.

For instance: If you need to know the area of a square with a side length of 3.5 feet and you have a table of values like Table A.1, it is clear that 3.5 is halfway between the 3 and the 4, for which values are given in the table. Therefore, the new value of the area will be about halfway between the 9 and the 16 in the table. A diagram like Figure A.2 illustrates the situation:

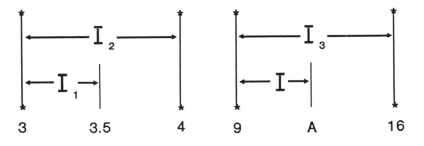

ORBITA 1991

Figure A.2—Developing an Interpolation Formula

In the left-hand part of the figure, the interval I_1 between the 3 and the new value 3.5 forms a fraction I_1/I_2 of the whole interval I_2.

In the right-hand part of the figure, the interval I between the 9 and the new value at A forms a fraction I/I_3 of the whole interval I_3.

This leads to the general interpolation formula:

$$\frac{I}{I_3} = \frac{I_1}{I_2}$$

It then follows that:

$$I = I_3 \times \frac{I_1}{I_2}$$

In this case:

$$I = \frac{(16-9)(3.5-3)}{(4-3)} = 7 \times \frac{.5}{1} = 3.5$$

The final answer is found by adding 3.5 to 9 to obtain 12.5. In this case the actual value could have been found more easily from the original equation $A = s \times s = 3.5^2 = 12.25$. But in many cases, when only the tabulated values are available, you can obtain an approximate answer by using the interpolation formula.

A.3—Working with Decimals and Rounding Numbers

Decimals represent tenths and powers of tenths of one. The way we express dollars and cents is a familiar example of the use of decimals. We know, for example, that $1.25 is the same as one dollar plus 25/100 dollars (because there are 100 cents in a dollar). Similarly, 2.4 represents two plus four-tenths. When using a calculator, it is easier to work with decimals than with fractions. For example, 3/5 is a simple fraction but, carrying out the division: Key [3], Key [/], Key [5], Key [=], results in the display of 0.6, which is the decimal equivalent of 3/5.

The best procedure when using a calculator is to convert all fractions to decimals.

⊨ Rounding Numbers

For most practical problems, it is unnecessary to be absolutely precise in making calculations. Although calculators can handle very large numbers and many decimal places, for every day use, the result of a calculation will be accurate enough if you "round" the answer to a couple of decimal places.

The usual rule that is usually used for rounding off is:

- When the digit to be dropped is five or more, increase the final digit by one;
- When the digit to be dropped is four or less, the final digit remains the same.

 EXAMPLE To round 1.006 to two decimal places, write 1.01; but, to round 1.004 to two decimal places, write 1.00.

A.4—Working with Powers of Numbers

When a number is multiplied by itself, it is said to be "squared" or raised to the second "power." This operation is written as y^2, meaning that the number represented by y is squared. Here, 2 is called the "power" or "exponent." In the same way, y^3 means y is raised to the third power ($y \times y \times y$), and so on.

Negative powers are used to indicate that the number should be inverted. For example, 10^{-1} represents 1/10; 10^{-2} represents 1/100; and so on.

Calculators have special keys for evaluating exponents. The [Y^x] key is used for this. For example, to evaluate 9^4, the entry sequence is: key [9], key [Y^x], key [4], key [=], and the answer displayed is 6,561.

Negative powers require the use of the "change-sign" key marked [+/–]. It is used to change the sign of the previous entry, which may be either a number or, as in this case, the power of a number. For example, to find the value of 9^{-4} which means $1/9^4$, the entry sequence is: key [9], key [Y^x], key [4], key [+/–], key [=], and the answer displayed is 0.0001524. (You may note that this is the reciprocal of the previous answer, 6,561.)

To find roots of numbers, the sequence is the same except that another key, labeled the "inverse" key, must also be used before the [Y^x] key. For example: To find the fourth root of 81—that is, the number which multiplied by itself three times will equal 81, the entry sequence is: key [81], key [Inv], key [Y^x], key [4], key [=], and the answer displayed is 3.

⇨ Dealing with Large and Small Numbers

In order to simplify operations involving either large or small numbers, a system of notation using powers of 10 is commonly used. For example, 1,000 can be written as 1×10^3, or as 10×10^2. In the same way, the decimal fraction 0.001 can be written as 1×10^{-3}, or alternatively as 10×10^{-4}. In fact, any number can be written in terms of powers of 10. Other examples are:

$$796{,}423 = 7.96423 \times 10^5 = 79.6423 \times 10^4$$

$$0.41865 = 4.1865 \times 10^{-1} = 41.685 \times 10^{-2}$$

In each case, the power of 10 indicates the number of places that the decimal point should be moved to the right or left to obtain the original number.

A.5—Using Logarithms

The reason logarithms were developed was to simplify calculations. Although they are seldom (if ever) used directly now, they are used extensively within calculators to simplify their internal operations. Today, calculators carry out the manipulations and present the results; you no longer need to consult tables of logarithms.

Nonetheless, it is sometimes useful to have a general understanding of their uses because they arise in many practical situations. In this handbook, for example, they arise in calculations concerning population growth (5.14, page 123); spread of disease (4.11, page 89); the pH scale of acidity (5.9, page 113, and 8.10, page 196); the Richter Scale of earthquake magnitudes (5.12, page 120); radioactive decay (9.6, page 237); and the exponential function (A.7, page 257).

In the previous section, dealing with powers of numbers, we noted, for example, that 10^4 means 10 raised to the 4th power and $10^4 = 10,000$. Now, if we consider the number 10,000, we can say that 4 is the power to which we must raise the "base" (in this case 10) in order to obtain the number, and 4 is called the "logarithm" of 10,000. That is: $10^{\log 10,000} = 10,000$—or, in general: $B^{\log N} = N$, where B is the base and N is the number.

Although B could be any number, in practice, only two base values are used:

1: $B = 10$ is called the base of common logarithms, and we have $10^{\log N} = N$.

2: $B = e$ is called the base of natural logarithms, and we have $e^{\ln N} = N$ where the symbol $\ln N$ is used to mean "logarithm to the base e" of N. Here e has the value $2.71828....$

Many calculators have two keys [log] and [ln$x$].

The relationships that make logarithms useful are as follows. They are written for logarithms to the base 10, but they also apply to natural logarithms.

$$\text{Log } (xy) = \log x + \log y \quad (xy \text{ is } x \text{ times } y)$$

$$\text{Log } (x/y) = \log x - \log y, \text{ and}$$

$$\text{Log } (x^n) = n \log x.$$

To illustrate the use of logarithms, we see in Section 5.14, page 123, that if a population Xt at a time t is growing at a rate r, it is related to the initial population Xo by the equation:

$$Xt = Xo \ e^{rt}$$

or, $Xt \div Xo = e^{rt}$.

If we wish to find the time at which the population is doubled, for example, when $Xt \div Xo = 2$, we have: $2 = e^{rt}$.

To solve this equation, we take the natural logarithm of each side and obtain: $\ln 2 = \ln e^{rt} = rt$, and using a calculator we can evaluate $\ln 2$ by entering: key [2], key [ln X], and obtain the answer: $\ln 2 = 0.693$.

Hence, from the equation we have $rt = 0.693$ and $t = 0.693 \div r$. (See Section 9.6, page 237, for a discussion of radioactive decay which can be analyzed in the same way.)

Antilogarithms are the inverse of logarithms, and in the previous example we can check the result by finding the antilogarithm of 0.693 by entering the following sequence in a calculator: key [0.693], key [inv], key [ln x].

The answer displayed will be 2, verifying that $\ln 2 = 0.693$.

A.6—The Algebra You Might Need

We mentioned the use of letters to represent numbers, which is the starting point of algebra, in A.1, page 243.

That section briefly explains what is meant by an algebraic equation and how to solve it, and outlines the general subject of functions and graphs.

Usually in practical problems, the first step is to formulate the problem in words; the second is to translate those words into mathematical terms to find a solution. It is here that algebra is of most use as a method of setting up equations that can then be solved by standard techniques. There are no precise rules for formulating problems in mathematical terms, but general guidelines to be followed include the following steps:

1: Identify the unknown quantities you are trying to find.

2: Represent one of these unknowns by a letter, say, x.

3: Specify the other unknowns in terms of x.

4: Analyze the problem carefully, and draw up one or more equations that express the known relationships.

5: Solve the equations.

6: Check the solution against the original statement of the problem.

The following example shows how you can use algebra in everyday use.

Q: A train traveling 80 mph is twenty miles from Grand Central Station at 8:45 AM. At what time will the train arrive at the station?

This everyday calculation may be figured using an algebraic equation. Since rate × time = distance, the equation may be set up as follows:

80 mph × time = 20 mi

The key to an algebraic equation, where a value or values are unknown, is that you may add, subtract, multiply, or divide by any known value, but you must do so to both sides of the equation, such that:

$$\frac{80 \text{ mph} \times \text{time}}{80 \text{ mph}} = \frac{20 \text{ mi}}{80 \text{ mph}}$$

and then,

time = 20/80
     = .25h
     = ¼ hour or 15 minutes.

A: The train will arrive at 9 AM.

▭ Graphs of Algebraic Equations

For many purposes, it will be helpful and informative to put an algebraic equation into graphic form. Here we take a general approach to graphs by explaining the use of rectangular coordinate axes. These axes are two straight lines that cross at right angles at a point called the "origin" of the axes. By convention, a scale of positive numbers is drawn along the horizontal axis to the right and another scale of positive numbers along the vertical axis upward. In each case, a corresponding scale of negative numbers is drawn along the axis in the opposite direction. Figure A.6A illustrates the axes and the scales.

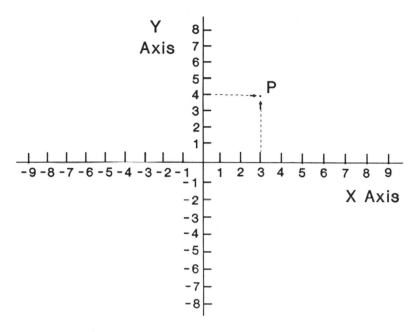

Figure A.6A—Rectangular Coordinate Axes

These axes are used to specify the location of a point in the diagram by expressing its distance to the right or left by a number on the horizontal scale, and its distance up or down by a number on the vertical scale. The horizontal axis is called the X-axis, and the vertical axis is called the Y-axis.

To locate a point P, its coordinates might be, for example, $X = 3$ and $Y = 4$, which means it is located 3 units to the right and 4 units up. This is written P (3,4). It is shown in Figure A.6A.

We can now plot the graph of an algebraic equation—for example, $Y = 2X + 3$.

The first step in preparing the graph is to draw up a table of values of X and Y. For example, for $X = 2$, we put this value in the equation and see that $Y = (2 \times 2) + 3 = 4 + 3 = 7$. In this way we obtain a table of corresponding values of X and Y.

$$X = \quad -1 \ 0 \ 1 \ 2 \ 3 \quad 4$$
$$Y = \quad \ \ 1 \ 3 \ 5 \ 7 \ 9 \ 11$$

We can now use these values as the coordinates of a set of points, which can be joined by a line to form a graph representing the function $Y = 2X + 3$. It is plotted in Figure A.6B.

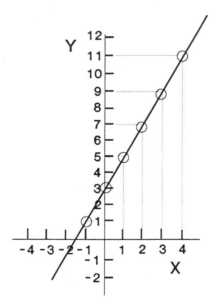

Figure A.6B—Plot of the Equation $Y = 2X + 3$

This method of plotting functions is generally useful for illustrating how one quantity varies in relation to another.

A.7—The Exponential Function and Its Many Uses

As explained in Section A.4, page 249, powers of numbers can be found using the [y^x] key on a calculator. For example, to find the value of 8^3, the entry sequence is: key [8], key [y^x], key [3], key [=], and the answer displayed will be 512.

The base of natural logarithms $e = 2.71828$ is introduced in A.5, page 251. Powers of e can be found using the [y^x] key of the calculator with the same entry sequence; some of the values of e^x and also e^{-x} found in this way are as follows:

X	=	0	1	2	3	4
e^x	=	1.00	2.718	7.39	20.09	54.6
e^{-x}	=	1.00	0.368	0.135	0.050	0.018

Figure A.7A shows a graph of the function $Y = e^x$ and Figure A.7B shows a graph of the corresponding function $Y = e^x$.

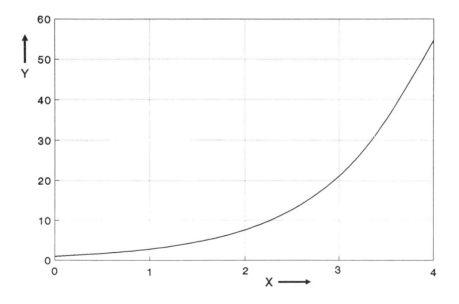

Figure A.7A—The Exponential Function $Y = e^x$

Fig. A.7B
The Negative Exponential Function $Y = e^{-x}$

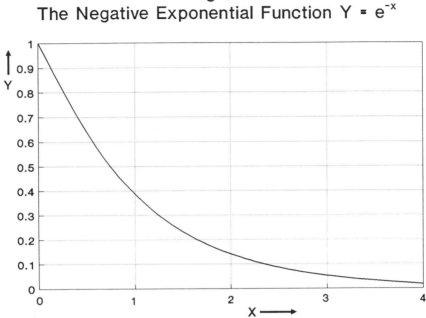

You can see in Figure A.7A that the value of Y increases rapidly as X increases, which gives meaning to the frequently heard remark that something "is increasing exponentially." Similarly, in Figure A.7B you can see that the value of Y decreases very rapidly as X increases—that is, Y decreases exponentially.

Four practical problems involving the exponential function are discussed in the text. The first three concern the increase in a money deposit when compound interest is added (see Section 3.1, page 43 and Section C.1, page 307); the spread of diseases such as AIDS (see Section 4.11, page 89); and the growth of human populations (see Section 5.14, page 123). The fourth practical problem concerns radioactive decay (see Section 9.6, page 237.)

In the first three cases, the amount Pt at time t is related to the initial amount Po by the equation $Pt = Po \times e^{rt}$ where r is the rate of growth. If this equation is plotted, it shows the rapid growth of a quantity that follows an exponential function. It is most convenient to plot the ratio $Pt \div Po$ against e^{rt}. A table of values is as follows, and Figure A.7C plots the function for two rates of growth: $r = 0.1$ and $r = 0.07$:

r	t	=	0	5	10	15	20	25	30	35
0.1	$e^{.1t}$	=	1.00	1.65	2.72	4.48	7.39	12.18	20.09	33.12
0.07	$e^{.07t}$	=	1.00	1.28	2.01	2.86	4.06	5.75	8.17	16.44

Applying this method of analysis to the growth of human populations illustrates its usefulness. (See Section 5.14, page 123).

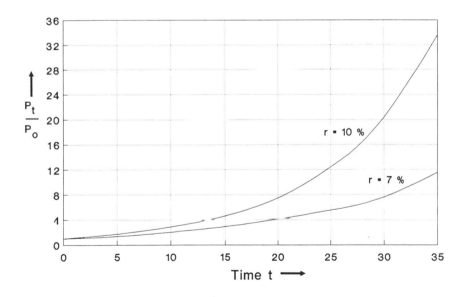

Therefore, if you want to determine how long it will take for a population to double, you can see that $Pt \div Po = 2$ when the population has doubled, and in that case the equation becomes: $2 = e^{rt}$. By taking logarithms of both sides, you find that: $\ln 2 = rt$, and from this: $rt = 0.693$ and, therefore, $t = 0.693 \div r$.

If the growth rate of a population is found to be 1% per year this relationship indicates that t is $0.693 \div 0.01 = 69.3$ years. The conclusion is that if a population is growing at 1% per year, it will double in a period of 69.3 years.

If the growth rate is 2% per year, the doubling time is about 35 years.

The same method can be used to work out tripling times and other cases. Table A.7 shows some results.

TABLE A.7

Doubling and Tripling Times for Population Growth

Annual Percent Increase	Doubling Time (years)	Tripling Time (years)	Annual Percent Increase	Doubling Time (years)	Tripling Time (years)
0.5	140	220	8.0	9	14
0.8	87	137	9.0	8	12
1.0	69	110	10	7	11
2.0	35	55	20	3.5	5.5
3.0	23	37	30	2.3	3.7
4.0	17	27	40	1.7	2.7
5.0	14	22	50	1.4	2.2
6.0	12	18	60	1.2	1.8
7.0	10	16	70	1.0	1.6

The same method, used with the negative exponential, gives the result that: $t = 0.693 \div r$ where r in this case is the rate of decay. In this way, the half-life of a radioactive element can be determined at the time during which the number of atoms has decayed to one-half of the original number.

A.8—The Geometry You Might Need

Plane geometry is the study of points, lines, and angles on a two-dimensional flat surface.

Some key terms in plane geometry are:

- ANGLE: a figure formed by two straight lines protruding from the same point
- side: a line that forms a boundary of a geometric figure
- vertex: a point at which two lines or curves intersect
- right angle: the angle formed by two lines that are perpendicular to each other

- LINE: a straight or curved element generated by a moving point and that extends only along the point's path
- perpendicular: a line that is at right angles to a line or plane
- parallel: when two lines extend in the same direction and are always equidistant from each other

- CIRCLE: a closed, plane curved figure all of whose points are equidistant from the center
- radius: the length of a line that begins anywhere on the circle and ends at the circle's center
- diameter: the length of a line that begins on the circle, crosses through the center, and ends on the circle
- circumference: the boundary of a circle

- PERIMETER: the boundary of a closed plane figure
- circumference (C): the perimeter of a circle; $C = \pi d$ (where d = diameter), or $C = 2\pi r$ (where r = radius)

- AREA (A): the surface within a set of lines
- key equations: square—$A = s^2$ (where s = side); rectangle—$A = l \times w$ (length $\times$ width); triangle—$A = \frac{1}{2}(b \times h)$ (where b = base and h = height); parallelogram—$A = b \times h$ (base $\times$ height); circle—$A = \pi r^2$ (r = radius)

- POLYGON: a closed, plane figure bounded by straight lines

- QUADRILATERAL: a four-sided, closed plane figure
- parallelogram: a quadrilateral with opposite sides equal and parallel
- rectangle: a parallelogram all of whose angles form right angles
- square: a rectangle that has four equal sides and four right angles

- TRIANGLE: a three-sided, closed plane figure
- right triangle: the result of two corners of a rectangle being connected with a line
- hypotenuse: the side of a right triangle that is opposite the right angle itself
- legs: the side of a right triangle that is *not* the hypotenuse

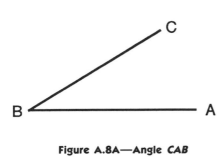

Figure A.8A—Angle *CAB*

When two straight lines, marked as *AB* and *BC* in the figure, meet at a point, they form an angle. If the line *BC* rotates about the point *B* and moves through a full circle, the angle measures 360 degrees. If *BC* rotates one-quarter of the full circle (a closed plane curve), or 90 degrees, the angle is a "right angle." Each degree can be subdivided into 60 minutes, and each minute can be subdivided into 60 seconds. In some cases, it is useful to measure angles in radians—which are defined by the relationship 2π radians $= 360$ degrees. (Therefore, there are approximately 57.32 degrees in a radian.) If you connect points *A* and *C*, the three-sided figure you form is a triangle, with three angles—which, by an important theorem of geometry, add up to 180 degrees.

Another important principle of geometry, the Pythagorean theorem, states that if one of the angles of a triangle is a right angle, the lengths of its sides are related as: $AC^2 = AB^2 + BC^2$. That is: The length of the side opposite the right angle squared is equal to the sum of the squared lengths of the other two sides.

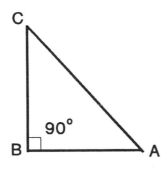

Figure A.8B—Right Triangle

In many practical problems, you must determine the perimeter (the distance around the sides of an object), or the area (the surface) of an object. For regular objects—such as triangles, rectangles, and circles—you can follow formulas to find the areas and perimeters. For an irregular object, it is often possible to divide it into a number of regular shapes for which formulas are available.

Some common formulas are as follows:

➤ Quadrilaterals

RECTANGLE

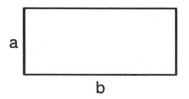

Area = base × height
= $b \times a$

SQUARE

If $a = b$, the rectangle is a square and the area is a^2.

Figure A.8C—Rectangle

➤ Triangles

RIGHT TRIANGLE

If two corners of a rectangle are connected with a line, two right triangles are formed. The area of each is half the area of the rectangle. That is area ($A = \frac{1}{2}$ base, multiplied by height, or $A = \frac{1}{2}$ $(b \times h)$ or $A = (b \times h) \div 2$.

OTHER TRIANGLES

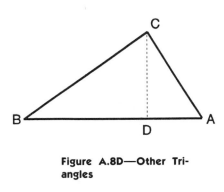

Other triangles can be divided into smaller right triangles, and you can find the area by adding the area of its parts. In this case: By drawing a dotted line *CD*, you form two right triangles, and you find the total area by adding the two areas.

Figure A.8D—Other Triangles

Two other triangles are equilateral triangle, which has all three sides equal, and isosceles triangle, which has just two sides equal.

Circle

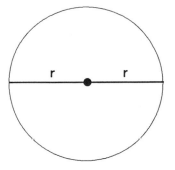

The diameter, by definition, is twice the radius. The quantity Pi (π) is defined by the equation π = circumference $\div$ diameter = 3.14159265.... The area is $A = \pi r^2$.

Figure A.8E—Circle

Three-dimensional Geometry

This type of geometry involves the study of points, lines, and angles for three-dimensional objects.

Some key terms in three-dimensional geometry are:

- RECTANGULAR SOLID: a rectangular figure that has three dimensions

- CUBE: a regular solide of six equal square sides

- CYLINDER: the surface traced by a straight line that is parallel to a fixed straight line and intersecting a fixed two-dimensional closed curve

- SPHERE: a solid that is bounded by a surface that consists of all points at a certain distance from the center

- VOLUME: the amount of space that occupies three-dimensional objects (measured in cubic dimensions)
- key equations: rectangular solid—$V = l \times w \times h$ (where l = length, w = width, and h = height); cube—$V = s^3$ (where s = side); circular cylinder—$V = \pi r^2 h$ (where r = radius and h = height); sphere—$V = 4/3\pi \; r^3$ (where r = radius)

- SURFACE AREA (A): the area of a three-dimensional region
- key equations: rectangular solid—$A = 2 \; (l \times w) + 2 \; (l \times h) + 2 \; (h \times w)$ (where l = length, w = width, and h = height); cube—$A = 6 \; (s^2)$ (where s = side); cylinder—$A = \pi \; r^2$ (where r = radius); and sphere—$A = 4\pi r^2$ (where r = radius)

Here are some common formulas in three-dimensional geometry:

✏️ Cylinder

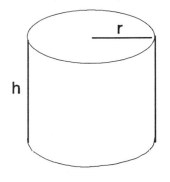

Total area of each of the circular ends $= \pi r^2$.

Total area of the cylindrical surface $= A = 2\pi\, rh$.

Volume $= V = \pi r^2 h$.

Figure A.8F—Cylinder

✏️ Sphere

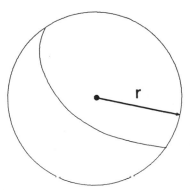

Area of the surface $= A = 4\pi r^2$.

Volume $= V = 4/3\ \pi r^3$.

Figure A.8G—Sphere

With these formulas, you can find the perimeter, area, or volume of many regular objects. For more complicated shapes, you will find additional formulas in reference texts. (In some cases, you can calculate approximate answers using the formulas given here.) If it is possible to immerse an irregular object in water, its volume can be found indirectly by measuring the volume of water that it displaces.

A.9—The Trigonometry You Might Need

The branch of geometry concerned with right triangles is called trigonometry.

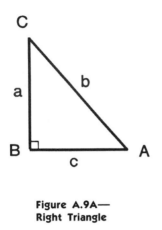

**Figure A.9A—
Right Triangle**

In the figure, the three angles may be denoted as angle A, angle B, and angle C; the three sides opposite these angles are marked as a, b, and c. Most of trigonometry is concerned with relationships involving six ratios of lines and angles. Each angle has a set of ratios. We express them here for the angle A, as follows: Sine $A = a \div b$; Cosine $A = c \div b$; and Tangent $A = a \div c$. (In each case, the reciprocal ratio has also been given a name, but they are seldom used in everyday calculations.)

Because these ratios are often used, many calculators have keys that give the values for any angle entered.

For example: If you key in 45 and press the sine key [sin], the display show the value 0.707. Another example: If you key in 150 and press the tangent key [tan], the answer is -0.577. In this case, the minus sign appears because of the conventions that have been adopted regarding directions along the coordinate axes. (See the discussion of coordinate axes in Section A.6, page 254.)

In many problems, it is useful to be able to "solve" triangles—that is, determine the unknown sides or angles when some sides and angles are known. For this purpose, probably the two most useful formulas that apply to all triangles, not just to right triangles, are as follows:

1: The Law of Sines:

$a \div \mathrm{Sin}\ A = b \div \mathrm{Sin}\ B = c \div \mathrm{Sin}\ C$

2: The Law of Cosines:

$a^2 = b^2 + c^2 - 2\ bc\ \mathrm{Cos}\ A$

Here the sides a, b, and c and the angles A, B, and C are as shown in the diagram above.

The Law of Cosines is given in the form that expresses the side marked a in terms of the other two sides. There are corresponding statements for the other two sides, as follows:

$b^2 = a^2 + c^2 - 2\ ac\ \mathrm{Cos}\ B$, and

$c^2 = a^2 + b^2 - 2\ ab\ \mathrm{Cos}\ C$

A.10—Elements of Probability Theory

Most of us are familiar with the general concept of probability. In general terms, it means the likelihood that something will happen. In mathematical terms, it can be defined as either the relative frequency of the occurrence of an event in a large number of trials, or the ratio of the number of occurrences of an event to the total number of equally likely cases.

For example: If a coin is tossed 1,000 times and "heads" turns up 500 times, then the probability of getting a "head" in a single toss is said to be 500/1,000, or 0.5 (50%).

👉 The Probability Scale

At one end of the probability scale, a value of 0 means that an event is impossible; there is no chance it will ever occur. At the other end of the scale, a probability of 1 means the event is certain to occur. Between these extremes are many possible outcomes, varying from very low values, if an event is extremely unlikely, up to values near 1, when we can be almost sure that it will occur. In tossing a coin, the probability that a head will come up is ½ or 0.5 or 50%; in rolling a die, the probability that a 6 will come up is 1/6 or 0.166 or 16.6%. The probability of precipitation is discussed in Section 5.5, page 106.

Having described the measuring scale, let's review the rules by which probabilities are analyzed. The two basic laws are the "addition" law and the "multiplication" law. They can be illustrated in relation to a situation where they can both apply—for example, at a racetrack, where you are considering different types of wagers.

👉 Addition Law

If you place a bet on each of two horses, A and B, that are running in the same race, and your bets are straight bets that the horse will win, then your probability that one of the two will win is the *sum* of the separate probabilities that A will win and the probability that B will win.

➩ Multiplication Law

If, on the other hand, you bet on horse A in the first race and then, if you win, you bet your winnings on horse B in the next race, your probability of winning both times is the *product* of the two probabilities (that is, the probability that horse A will win in the first race multiplied by the probability that horse B will win in the second race).

These simple laws cover many of the common situations in which probabilities play a part. They are basic to the development of many of the probabilities in games and sports described in Chapters 6 and 7.

They also underlie the aspect of probability theory concerned with permutations and combinations, which we review next.

➩ Permutations and Combinations

We can say that the three letters ABC form a *combination* of letters; and we can go on to identify as *permutations* the six different arrangements of the order of these letters—that is, ABC, ACB, BCA, BAC, CAB, and CBA.

This example shows what we mean by combinations and permutations. In the first case, the order in which the letters are arranged does not matter; in the second case, it does.

The significance of the difference between the two can be seen by considering a practical example:

If there are nine horses in a race, you can bet that a certain three of the nine will be the fastest horses in the race without identifying in what order the three will finish. In this case, you are concerned with one *combination;* the order of the three horses does not matter. However, if you bet not only that the three horses will place in the first three, but also that they will finish in a particular order, you are betting on only one *permutation* of the six that are possible.

Number of Permutations

If you have a combination of things that are all different, the number of permutations can be calculated as follows:

 Consider the three letters A, B, and C. For the first position in the permutation, there are three possibilities—A, B, or C. When that position has been filled, there are only two possible choices for the second position. And finally, for the third position, there is only one possible choice. Therefore, the total number of choices will be $3 \times 2 \times 1 = 6$.

As a shorthand notation, the product $3 \times 2 \times 1$ is called "factorial 3" and is written "3!"

The number of permutations is expressed by another shorthand notation. For example: 4P3 is a short way of writing "the number of permutations of four things taken three at a time." In this type of notation, the number of permutations of the three letters A, B, and C taken three at a time, as we outlined above, would be written 3P3 = 3! = 6.

In general, we can write $n\mathrm{P}n = n!$ for the case in which all items in a combination are different. However, if some of the items are the same, the number of permutations is reduced; the general formula is that if there are n things including p things of one type, q things of another, and r things of a third type, then the number of permutations $= n! \div (p! \times q! \times R!)$.

EXAMPLE The number of permutations in a deck of 52 cards with four suits is $52! \div (13! \times 13! \times 13! \times 13!)$.

☞ Number of Combinations

As we have seen, one combination can have many permutations. We showed that a combination of r things can have $r!$ permutations, which leads directly to the statement that the total number of permutations is the number of combinations multiplied by the number of permutations within each combination, that is, in shorthand: $n\mathrm{P}r = n\mathrm{C}r \times r!$ or $n\mathrm{C}r = n\mathrm{P}r \div r!$

☞ Applications of Permutations and Combinations

To illustrate the uses of these concepts, we can apply them to the calculation of probabilities of obtaining certain distributions of cards in the game of bridge. (See Section 6.8, page 143.)

EXAMPLE The total number of different hands of thirteen cards from a deck of 52 cards is written 52C13; the number of hands with, say, five spades out of the thirteen spades, is written as 13C5; hence, the chance that a hand will have five spades is 13C5 ÷ 52C13. If we wish to calculate the probability that a hand will have, say, five cards of one suit, four cards of another suit, three cards of a third suit, and one card of the fourth suit—that is, P(5–4–3–1)— we can write this probability as: 4! × (13C5 ×

13C4 × 13C3 × 13C1) ÷ 52C13. (The 4! factor is needed in the expression because the
suits can be any one of four—and therefore
there are 4! different ways in which the distributions with respect to the suits can occur. See
the paragraphs above concerning the number of
permutations.)

By substituting values for the terms in the expression, it
can be evaluated to be 0.1293. This probability and others
like it are listed in Section 6.8, page 143.

Odds and Odds Ratios

The terms "odds" and "odds ratios" are frequently used in
connection with gambling and betting. The probability that one thing
will happen rather than another is referred to as the odds, and the
ratio of the probabilities that two events will occur is called the odds
ratio.

EXAMPLE

If you carry out a series of trials and there
are four successes and six failures, the probability of success is 4 out of 10, or 0.4, and
the probability of failure is 6 out of 10, or
0.6. In betting parlance, this relationship would
be stated as the odds are "4 to 6 for" the
odds ratio would be 0.4 ÷ 0.6, which is the
same as 4/6 or 2/3.

If you were at the track and ten races were being run,
four would result in a win for you and six would result in a
loss. If you bet $2 on each race, your total outlay would be
$20, and you would have to win $5 on each of the four
successes to get back your $20 outlay. Therefore, you
would hope to break even if you accepted odds of "5 to 2
for"—that is, a $5 return for a $2 stake.

A.11—Basic Statistics, Polls and Surveys

To understand large bodies of data, you must reduce their complexity by determining certain measures, called "measures of central tendency," which serve to summarize their main features.

When data are arranged in order of magnitude, they often cluster around some central value that provides a useful guide to the meaning of the data.

Four measures are used to indicate different kinds of central values. They are the mean or average value, the median value, the mode, and the geometric mean.

✏ The Mean or Average

This is the value we usually mean when we speak of the "average." More correctly, it is the arithmetic average—because it is found by adding up all the figures and dividing by the number of figures in the set of data. For example: The average of 2, 4, and 6 is $(2 + 4 + 6) \div 3$, or 4.

The mean is fairly easy to calculate unless there is a great number of figures to be averaged. Its major drawback is that it can give a distorted picture of the central value if there are even a few very large or very small values in the set. For example: If there is one very rich person in a community, the mean of the incomes in the community will be influenced by the one extreme value so that it will not be a good measure of the truly "average" income.

✏ The Median

If the figures are arranged in numerical order, the middle value in the set is called the median. If there is a small number of observations and the number is even, it may be necessary to add the two middle numbers together and divide by two to obtain a value for the median. Otherwise, there would be more numbers on one side of the median than on the other.

The following set of data is obtained: 26, 29, 31, 35, 36, 39.

In this case, there is no middle value; to find a median, you must add the two middle values together and divide by 2—that is, $(31 + 35) \div 2 = 33$. This is called the median even though it is not strictly a member of the set.

 The median can be the same for different sets of data, such as the following: In the set, 22, 23, 24, 25, 26, 27, and 28, the median is 25; and in the set, 22, 23, 24, 25, 26, 27, and 100, the median is also 25.

Although the medians are the same, the two sets of data have at least one very different value.

On the other hand, this lack of sensitivity to extreme values can sometimes produce a more insightful result.

The Mode

This is the value that occurs most frequently in a set of data. In certain sets, there may be one value that occurs so often that it is clearly the most representative of the set, even if the mean and the mode are different.

Here's a case when the mode is the most useful. A manager of a store records the shirt-collar sizes sold over a certain period. At re-order time, he or she will want to order the size that sold most frequently—that is, the mode.

The Geometric Mean

The geometric mean is found by taking the nth root of the product of the n observations.

If four observations are taken with values 2, 4, 6, and 8, their geometric mean is $(2 \times 4 \times 6 \times 8)^{1/4} = 4.427$. (For comparison, the arithmetic mean is 5, the median is also 5, and there is no mode.)

The geometric mean is difficult to compute if there are many observations in the set, and it cannot be used if any of the values is negative or zero. It can be used as in the following example:

 Section 5.14, page 123, discusses the population growth at different growth rates. The principal feature of such growth is that the rate is a percentage of the population—and, therefore, it changes as the population changes. Since census figures are taken only once a decade, it is often necessary to estimate what the population was in the years between the measurements. In this case, the geometric mean is the best method. If the population of a city is measured as 10,000 in 1981, and 20,000 in 1991, the geometric mean is: $(10,000 \times 20,000)^{1/2} = 14,140$. For comparison, the arithmetic mean in this case is: $(10,000 + 20,000) \div 2 = 15,000$.

If the equation given in Section A.7, page 257, is used to calculate the population in 1986 (midway between the census dates), the geometric mean will be more accurate than the arithmetic mean.

There are many situations that involve either growing or decaying at a rate proportional to the amount present. These situations are described by an exponential function. (See Section A.7, page 257.) In these cases, the geometric mean is the best measure to use.

Frequency Distributions

When dealing with a large mass of data, you'll often find it helpful to arrange the data in groups or classes. The number of observations in a class is then called the "class frequency." The use of groups or classes can be illustrated with an example.

 EXAMPLE The marks obtained by 25 students in an examination are shown in Table A.11A. The marks are divided into classes, each five percentage points wide—that is, 76 to 80; 81 to 85; and so on.

The class frequencies are found and tabulated. After that, you can work with the frequency distribution rather than with the actual marks. This is done in calculating the arithmetic mean, although in this simple case it would not have been necessary.

You calculate the arithmetic mean by dividing the total of the products by the total number of marks—that is; 2,172.5 ÷ 25 = 86.9. It is a useful measure in this case if, for instance, you want to compare the performance of this group of students with the performance of another group taking the same examination.

TABLE A.11A

Frequency Distribution of Exam Results

Classes of Marks	Frequency in each Class	Central Class Value	Product of Frequency and Central Class Value	
96 to 100	3	97.5	3 × 97.5 =	292.5
91 to 95	5	92.5	5 × 92.5 =	462.5
86 to 90	7	87.5	7 × 87.5 =	612.5
81 to 85	6	82.5	6 × 82.5 =	495.0
76 to 80	4	77.5	4 × 77.5 =	310.0
TOTAL	25		TOTAL	2,172.5

🖊 Dispersion

Measures of central tendency do not give any indication how much spread or variation there is in the data, which is another very important factor when analyzing a set of observations.

There are several measures of dispersion, but the one that is most widely used is the "standard deviation."

🖊 Standard Deviation

The first step in calculating the standard deviation in a set of data is to find the arithmetic mean of the set. Then the deviations of all of the observations from the mean are found. These deviations are squared and added together into a sum of squares. The final step is to take the square root of this sum.

The definition of the standard deviation is the square root of the mean of the squares of all the individual deviations from the mean of the set of observations or the frequency distribution.

The set of examination marks used above will provide an illustrative example for finding a standard deviation. As before, the arithmetic mean is calculated as 86.9.

Table A.11B gives results of the steps in the calculation.

TABLE A.11B

Calculating a Standard Deviation

Central Class Value	Frequency in the Class	Deviation from the Mean	Deviation Squared	Product of (deviation)^2 and frequency
97.5	3	97.5 − 86.9 = 10.6	112.4	337
92.5	5	92.5 − 86.9 = 5.6	31.4	157
87.5	7	87.5 − 86.9 = 0.6	0.4	3
82.5	6	82.5 − 86.9 = −4.4	19.4	116
77.5	4	77.5 − 86.9 = −9.4	88.4	353
TOTAL	25			TOTAL

Standard Deviation $= (966/25)^{1/2} = 6.22$.

For many distributions, the interval one standard deviation on each side of the mean contains about 70% of the observations. In the example, this interval is:

$$\text{from } 86.9 - 6.2 = 80.7$$

$$\text{to } 86.9 + 6.2 = 93.1$$

and it appears that $5 + 7 + 6 = 18$ of the 25 observations, meaning that 72% lie within this interval.

⬭ Coefficient of Variation

A useful measure of the relative variability of data is the "coefficient of variation," which is the standard deviation divided by the mean, expressed as a percentage. In the example, its value is $6.22/86.9 \times 100 = 7\%$.

⬭ Polls and Surveys

Poll or survey results are often stated in the following way: "Among those replying, Choice A was favored by 55%, and Choice B was favored by 45%. The poll of 1,000 adults gives results that are accurate within 5%, 19 times out of 20."

This means that if the poll were repeated twenty times, you could expect that in one case, the number favoring Choice A would lie outside a range extending from $(55 - 5)\%$ of 1,000 at the lower end to $(55 + 5)\%$ of 1,000—that is, from 500 to 600. In the other 19 repetitions of the poll, the results would lie within this range. For those favoring Choice B, the range is from 40% to 50%—that is, from 400 to 500.

Sometimes the results are said to be accurate to a certain percentage "95% of the time," instead of "19 times out of 20." The meaning is the same in these two cases.

The results of polls and surveys must be treated with care, particularly if there is an overlap in the ranges for the different choices. Even if there is no ambiguity in the questions asked and the statistics are valid, they must be examined to see if they support firm conclusions or, on the other hand, indicate that there is little to choose between the options presented.

A.12—Sequences and Series

In certain applications, successions of numbers or terms occur. These can be either sequences, if formed in accordance to some definite rule, or series, if formed in some other way.

A typical sequence is the set of numbers 1, 3, 5, 7, etc.—formed by adding 2 to each succeeding term.

A typical series is the sum of the numbers in the preceding sequence—that is, $1 + 3 + 5 + 7 +$ etc.

Two sequences and one series are of particular interest.

▱ Arithmetic Progressions

An "arithmetic sequence" (or "arithmetic progression") is one in which the successive terms are found by adding a constant to, or subtracting a constant from each successive term.

If we denote the first term by F, the last term by L, the constant difference between terms by d, and the number of terms by n, we can express the sum of the terms S using the following formulas:

$$L = F + (n - 1)\ d, \text{ and}$$

$$S = n\ (F + L) \div 2$$
$$= n[2F + (n - 1)\ d] \div 2$$

▱ Geometric Progressions

A "geometric sequence" (or "geometric progression") is one in which the successive terms are formed by multiplying or dividing by a constant.

With the same symbols as before except that r, the common factor, replaces d, the common difference, we have the following formulas for the geometric progression:

$$F,\ Fr,\ Fr^2,\ Fr^3, \text{ etc.:}$$

$L = Fr^{n-1}$ and

$S = Fr^n \div (r - 1)$

Note that a population grows according to a geometric progression if the rate of growth is a constant. (See Section A.7, page 257.)

➯ Exponential Series

The base of natural logarithms, e, can be expressed as the sum of an infinite series: $e = 1 + (1 \div 1!) + (2 \div 2!) + (3 \div 3!) + \ldots = 2.71828\ldots$ (See Section A.5, page 251.)

A.13—Computer Basics

The calculations in this book can be carried out without using a computer, though in some cases the printed results can be extended through the use of a personal computer and the appropriate software.

It is not the purpose of this section to explain how a computer should be used for solving the problems in the book, but it may be useful to include a very brief introduction to the subject for those who are not "computer literate."

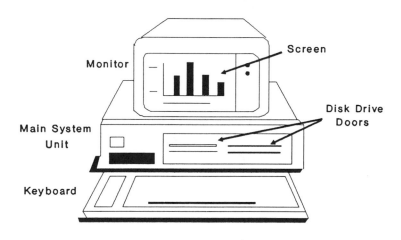

Figure A.13A—A Typical Desktop Computer

Probably the most basic questions concern what is meant by bits and bytes and the size of the memory in a computer. The electronic memory in computers is made up of a very large number of microscopic binary circuits, each of which exhibits only one of two possible states. Each may be thought of as an electronic switch that is either on or off. These states can be represented by the digits 1 and 0, which are called the "binary digits."

In technical terms, there are 8 "binary digits" or "bits" in one byte. The eight circuits that store one byte of information in memory can represent any one of 256 unique codes. This is so because there are 256 ways (that is, 2^8) that eight binary states can be uniquely combined. By convention, these codes are numbered 0 to 255, and it is the meanings that are ascribed to these codes that define the way in which the computer both responds to commands and stores information.

The first 32 codes (numbers 0 to 31) govern the way the computer responds to commands. These are called the "control codes"—and, unhappily, different brands of computers differ in the way they interpret these commands. The next 96 codes (numbers 32 to 127) represent information characters, including the numbers 0 to 9 and the letters A to Z and a to z, as well as many other common characters such as $, %, +, and the ubiquitous "space" that marks the end of words. Happily, most computers observe the standard set by the American Standard Code for Information Interchange (ASCII) for these 96 codes. The high-order codes (numbers 128 to 255) are used in a variety of ways by various computers or programs to either extend their control command set or supplement the characters available in the ASCII standard character set.

A typical personal computer today may have a dynamic memory of one megabyte (that is, 1,024 kilobytes, or 1,048,576 bytes). Not all of this is available for the use of programs and data, however. The computer itself claims about 37.5% for internal manipulation and control functions; an additional 2 to 12% is required by a special control program called the operating system. Application programs such as word processors, spreadsheets, and data base managers vary widely in the amount of dynamic memory that they demand. Simple programs may use only about 32 kilobytes, but many modern programs gobble up 256 kilobytes and more of the remaining dynamic memory.

A printed page of single-spaced text has about 700 words—or about 4,000 characters, including spaces. Therefore, a computer with 512 kilobytes of usable dynamic memory can store about 524,288 ÷ 4,000 = 130 pages of text.

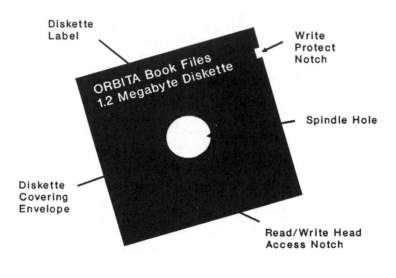

Figure A.13B—A 5¼-inch Diskette

In addition to the dynamic memory within the computer, the computer must have external memory storage—because when the computer is shut down, all the information stored in its dynamic memory dies. Magnetic disks are widely used to store programs and information that are required again. Many modern computers have permanent disk installations called hard-disk drives for this purpose. Typically, they store from 40 to more 160 megabytes. Simpler computers rely upon removable diskettes. The original standard 5¼-inch diskettes held about 360 kilobytes. Modern 5¼-inch diskettes hold about 1.2 megabytes of data, while the more compact 3½-inch diskettes hold either 720 kilobytes or 1.44 megabytes.

Removable diskettes are inserted into a reader, called a disk drive, which forms part of the computer system. Part of the data from the disk is transferred into the internal memory in the computer, where it can be processed. For example, if it is text, it can be edited and then sent to a printer, which also forms part of the computer system. The complete text of this book with tables and graphs can be stored on two 360-kilobyte diskettes or a single 720-kilobyte, 1.2-megabyte or 1.44-megabyte diskette.

Appendix B

Getting the Units of Measurement Right

B
ecause units of measurement come into most numerical problems, you need to understand those units to make accurate calculations. The first section of this appendix provides a glossary of units, arranged alphabetically, that will explain some of the units that are unfamiliar to many people. In many cases, we have defined the units where they come into the text; if so, there is only a page reference in the listing.

B.1—Units of Measurement from A to Z

A

Acre ....Area of 43,560 square feet, in a shape (for
example, approximately 209 feet square);
640 acres = 1 square mile or section.
Ampere (Amp) ...Electric current. *See* 8.6, page 182.
Angstrom ....Wavelength equal to 10^{-10} meters.
Astronomical
Unit ....Distance. *See* 9.2, page 214.
Atmosphere ....Pressure. *See* 5.1, page 99.

B

Bar ....Pressure in the atmosphere. *See* Kilopascal.
Barrel ....Volume equal to 42 gallons.
Beaufort Scale ....Wind velocity. Numbers from 0 (calm), 2
(light breeze), 5 (moderate breeze), 8 (gale),
up to 12 (hurricane).
Bel ....*See* Decibel.
Binary system of
numbers ....*See* A.13, page 285.
Bit ....Abbreviation for binary digits. *See* A.13, page
285.
Board foot ....Measure of lumber. *See* 8.9, page 191.
British Thermal
Unit (BTU) .....Amount of heat needed to raise the temper-
ature of one pound of water one degree
Fahrenheit. Equals .252 calories. *See* 8.8,
page 189.
Bushel ....Volume used in dry measure equal to eight
gallons.
Byte ....Equals eight Bits. *See* A.13, page 285.

C

Calorie ....Amount of heat needed to raise four pounds of water one degree Fahrenheit. Equals the metric calorie, which is the heat needed to raise one kilogram of water one degree Celsius. *See* 4.8, page 83.

Carat ....Weight used to measure precious stones. Equals 200 milligrams.

Chain ....Length used in land surveying. Equals 66 feet.

Cord ....Volume of wood. The standard cord is four feet high, four feet wide, and eight feet long. Another unit is the "face" cord, which can be either four or eight feet long, four feet high, and either twelve, sixteen, or eighteen inches wide.

Coulomb ....Electrical charge. A metric unit.

Cup ....Volume equal to 8 fluid ounces. 2 cups = 1 pint.

D

Decibel ....Relative intensity of sounds or electrical powers. The ear can just detect a change in loudness when the power level changes by 1 dB. Dynamic range of the ear may extend from 1 dB to 120 dB. Also expresses gain in an electronic circuit by relating voltage in and out—for example, doubling voltage means there is a 6 dB gain.

Degree of angle ..*See* A.8, page 263. 1 degree = 60 minutes = 3,600 seconds.

Degree Celsius ...Temperature. *See* 5.2, page 102.

Degree
   Fahrenheit ......Temperature. *See* 5.2, page 102.

Degree of Lati-
 tude ....Measure of distance from Equator. Equator
 is 0 degrees; North Pole is 90 degrees
 North. 1 degree = 60 nautical miles = 60
 minutes = 3,600 seconds.

Degree of Longi-
 tude ....Measure of distance east or west of the
 Prime Meridian through Greenwich, England.

Dyne ....Force—a metric unit. 100,000 dynes = 1
 newton.

E

Erg ....Work or energy—a metric unit. 10 million
 ergs = 1 joule.

F

Fathom ....Measure of water depths equal to six feet.
Fluid ounce ....Liquid measure. 8 fluid ounces = 1 cup =
 ½ pint.
Furlong ....Length equal to 660 feet or 220 yards.

G

Gallon ....Liquid measure equal to 231 cubic inches or
 four quarts.
Gram ....Mass or weight—A metric unit.
Gross ....12 dozen, or 144.

H–J

Hectare ....Area—a metric unit equal to 2½ acres.
Hertz ....Frequency equal to one cycle per second.
Horsepower ....Power required to raise 550 pounds one foot
 in one second. Equals 746 watts. *See* 8.15,
 page 204.
Hundredweight ...Weight equal to 100 pounds.
Joule ....Work or energy—a metric unit. Equals 10
 million ergs.

K

Karat ....Proportion of gold in an alloy. 1 karat means that 1/24th of the alloy is gold.

Kilogram ....Mass or weight—a metric unit equaling 2.2 lbs.

Kilopascal ....Pressure—a metric unit used in meteorology. 100 kilopascals = 1000 millibars. 101.3 kilopascals = 1 atmosphere.

Knot ....Speed of one nautical mile per hour.

L

Light year ....Astronomical distance. *See* 9.2, page 216.

M

Meter ....Length—a metric unit. Equal to 39.37 inches.

Micron ....Length—a metric unit. 1 million microns = 1 meter.

Mil ....Length equal to one thousandth of an inch.

Millibar ....Pressure in the atmosphere. *See* Kilopascal.

Minute of
  angle ....*See* Degree of angle.

Minute of latitude
  and longitude ...*See* Degree of latitude and longitude.

N

Nautical mile .....Length equal to 1.15 miles or one minute latitude.

Newton ....Force—a metric unit.

O–P

Ohm ....Electrical resistance. *See* 8.6, page 182.

Ounce ....Weight equal to 1/16th of a pound.

Parsec ....Distance. *See* 9.2, page 215.

Pascal ....Pressure. *See* Kilopascal.

Peck ....Volume used in dry measure. 4 pecks = 1 bushel.

pH Scale ....Acidity. *See* 5.9, page 113.

Pi (π) ....Ratio of circumference of a circle to its diameter. *See* A.8, page 265.

Pound ....Mass or weight. *See* 9.3, page 217.

Proof ....Alcoholic content. Twice the percentage by volume of alcohol present--for example, 90 proof whiskey is 45% alcohol.

R

Radian ....Angular measure. Pi radians = 180 degrees. *See* A.8, page 263.

S

Second of
 angle ....*See* Degree of angle.

Second of latitude
 and longitude ...*See* degree of latitude and longitude.

Section ....Area used in land measure equal to one square mile.

T

Tablespoon ....Volume equal to three teaspoons or 0.9 cubic inches.

Teaspoon ....Volume equal to one-third tablespoon or 0.3 cubic inches.

Therm ....Quantity of heat. *See* 8.7, page 186.

Ton ....Weight equal to 2,000 pounds.

Tonne ....Weight—a metric unit equal to 2,200 pounds.

V

Volt ....Electrical potential. *See* 8.6, page 182.

W

Watt ....Electrical power. *See* 8.6, page 182.

Z

Zulu Time ....24-hour clock—for example, 1330 hours = 1:30 PM

B.2—The Metric System—Logical Units

Many nations have adopted metric units or are moving toward adopting them. Within the next ten years, the use of older systems of measurement, including U.S. weights and measures is expected to decline. A worldwide standard system of units would reduce the calculations required to convert between systems of units.

Because all metric units are based on multiples of 10, they are much easier to use than other systems. Units of length can be used to illustrate how metric units are built:

The basic unit of length is the meter. Adding the prefix "centi" (meaning one one-hundredth) produces a unit called the centimeter (cm)—100 centimeters equals one meter. Adding the prefix "milli" (meaning one one-thousandth) produces a unit called the millimeter (mm)—1,000 millimeters equals one meter. Adding the prefix "kilo" (meaning 1,000) produces a unit called the kilometer (km)—one kilometer equals 1,000 meters.

Several other prefixes are also commonly used, including "nano" (10^{-9}), "micro" (10^{-6}) and "mega" (10^{6}). Others, rarely employed, are "hecto" (10^{2}), "deca" (10^{1}), and "deci" (10^{-1}).

In addition to the units of length just described, the most frequently used metric units are as follows:

AREA

- Square meter (m^2)
- Hectare = 1 square hectometer (hm^2) = 100 meters × 100 meters

VOLUME

- Cubic meters (m^3)
- Cubic centimeter (cm^3) (or milliliter) (ml)
- Liters (*l*) (1,000 cubic centimeters)

MASS

- Gram (g)

- Kilogram (kg)
- Milligram (mg)
- Metric tonne (t) (1,000 kilograms)

TEMPERATURE

- Degrees Celsius (on the Celsius scale, zero degrees is the freezing point of water, 37 degrees is the normal body temperature; and 100 degrees is the boiling point of water.)
- Degrees Kelvin (on the Kelvin scale, -273.15 degrees Celsius is the freezing point of water.)

ANGLE

- Degree (360 degrees is a full circle)
- Radian (57.3 degrees; 2 π radians)

The following units of measurement relate to the meter-kilogram-second system (mks). This system uses meter as the unit of length, kilogram as the unit of mass, and mean solar second as the unit of time.

FORCE

- Newton (a unit of force that when applied to a mass of one kilogram gives it an acceleration of one meter per second per second)

PRESSURE

- Pascal (a unit of pressure resulting when the force of one newton is applied to an area of one square meter)
- Kilopascal (1,000 pascals) (units commonly used for tire pressures)

MECHANICAL ENERGY

- Joule (a unit of work done when a force of one newton moves a distance of one meter)

CURRENT

- Ampere (one ampere flowing in two straight parallel conditions of infinite length one meter apart in a vacuum produces a force between them of 2×10^{-7} newtons per meter length of conductors.) Also, one ampere = mks unit of electric current = one coulomb / sec

CHARGE

- Coulomb (one coulomb is the quantity of electricity that flows in one second when the current is one ampere)

POWER

- Watt (one watt is the amount of work done at the rate of one absolute joule per second)

ELECTRICAL ENERGY

- Kilowatt = 1000 watts; Kilowatt-hour (one kilowatt-hour is the energy used when 1 kilowatt of power is used for one hour)

ELECTRICAL POTENTIAL DIFFERENCE

- Volt (one volt is the potential difference that will cause one ampere to flow when the power dissipated is one watt)

ELECTRICAL RESISTANCE

- Ohm (one ohm is the resistance of a circuit when a potential difference of one volt causes a current of one ampere to flow)

FREQUENCY

- Hertz (Hz) (one hertz is equal to one cycle per second)

B.3—Conversions Between Units

Within the metric system, conversion between units is simple because factors of 10 are involved in most cases. However, because earlier systems of units are still in use, you will often need to convert between units related by more-complicated factors. Table B.3 lists many of the common factors and relationships between units.

An example will illustrate the use of the table. Under units of length, the table shows that 1 kilometer = 0.621 statute miles. Therefore: 8 kilometers = 8 × 0.621 = 4.968 miles. (In common usage, 8 kilometers = 5 miles.) Conversely, to convert in the opposite direction, the table shows that 1 mile = 1.609 kilometers. Therefore, 5 miles = 5 × 1.609 = 8.045 kilometers.

Another example: In an earlier system of units, tire pressures were measured in pounds per square inch. In the metric system, they are measured in kilopascals. The table shows that 1 pound per square inch = 6.896 kilopascals, or approximately 7 kilopascals. To convert in the opposite direction, the table shows that 1 kilopascal = 0.145 pounds per square inch. Therefore, 100 kilopascals = 14.5 pounds per square inch, which is the approximate atmospheric pressure at sea level.

TABLE B.3

Conversion Factors Between Units

Abbreviations

atm.	—	atmosphere(s)
cm.	—	centimeter(s)
cu.	—	cubic
cwt.	—	hundredweight
ft.	—	foot(feet)
fl.	—	fluid
gal.	—	gallon(s)
g.	—	gram(s)
hr.	—	hour(s)
Hg.	—	mercury
imp.	—	imperial
in.	—	inch(es)

kg.   —   kilogram(s)
km.   —   kilometer(s)
kpa.   —   kilopascal(s)
lb.   —   pound(s)
ltr.   —   liter(s)
m.   —   meter(s)
min.   —   minute(s)
mi.   —   mile(s)
ml.   —   milliliter(s)
oz.   —   ounce(s)
pa.   —   pascal(s)
pt.   —   pint(s)
qt.   —   quart(s)
sec.   —   second(s)
sq.   —   square
yd.   —   yard(s)

LENGTH

1 m.   = 100 cm.; 1 cm. = 10^{-2} m.
       = 39.37 in.; 1 in. = 0.025 m.
       = 3.281 ft.; 1 ft. = 0.305 m.
       = 1.094 yd.; 1 yd. = 0.914 m.
1 km.   = 0.621 mi.(statute); 1 mi. = 1.609 km.
1 cm.   = 0.394 in.; 1 in. = 2.540 cm.
       = 0.033 ft.; 1 ft. = 30.48 cm.
       = 0.011 yd.; 1 yd. = 91.44 cm.
1 in.   = 0.083 ft.; 1 ft. = 12 in.
       = 0.028 yd.; 1 yd. = 36 in.
1 ft.   = 0.333 yd.; 1 yd. = 3 ft.
       = 1.893 $\times$ 10^{-4} mi.; 1 mi. = 5280 ft.
1 yd.   = 5.681 $\times$ 10^{-4} mi.; 1 mi. = 1760 yd.

AREA

1 sq. m.   = 10^{-4} sq. cm.; 1 sq. cm. = 10^{-4} sq. m.
          = 1550 sq. in.; 1 sq. in. = 6.452 $\times$ 10^{-4} sq. m.
          = 10.76 sq. ft.; 1 sq. ft. = 9.290 $\times$ 10^{-2} sq. m.
          = 1.196 sq. yd.; 1 sq. yd. = 0.836 sq. m.
          = 3.86 $\times$ 10^{-7} sq. mi.; 1 sq. mi. = 2.590 $\times$ 10^{6} sq. mi.
          = 2.471 $\times$ 10^{-4} acre; 1 acre = 4,047 sq. m.
          = 10^{-4} hectares; 1 hectare = 10^{4} sq. m.
1 sq. cm.   = 0.155 sq. in.; 1 sq. in. = 6.452 sq. cm.
          = 1.076 $\times$ 10^{-3} sq. ft.; 1 sq. ft. = 929.3 sq. cm.
          = 1.196 $\times$ 10^{-4} sq. yds.; 1 sq. yd. = 8,361 sq. cm.

1 sq. in.	= 6.944 $\times$ 10⁻³ sq. ft; 1 sq. ft. = 144 sq. in.
	= 7.716 $\times$ 10⁻⁴ sq. yd.; 1 sq. yd. = 1,296 sq. in.
1 sq. ft.	= 0.111 sq. yd.; 1 sq. yd. = 9 sq. ft.
	= 3.587 $\times$ 10⁻⁸ sq. mi.; 1 sq. mi. = 2.788 $\times$ 10⁷ sq. ft.
	= 2.296 $\times$ 10⁻⁵ acres; 1 acre = 4.356 $\times$ 10⁴ sq. ft.
	= 9.291 $\times$ 10⁻⁶ hectares; 1 hectare = 1.076 $\times$ 10⁵ sq. ft.
1 sq. yd.	= 3.228 $\times$ 10⁻⁷ sq. mi.; 1 sq. mi. = 3.098 $\times$ 10⁶ sq. yd.
	= 2.066 $\times$ 10⁻³ acres; 1 acre = 4,840 sq. yd.
	= 8.361 $\times$ 10⁻⁵ hectares; 1 hectare = 1.196 $\times$ 10⁴ sq. yd.
1 sq. mi.	= 640 acres; 1 acre = 1.562 $\times$ 10⁻³ sq. mi.
	= 259 hectares; 1 hectare = 3.861 $\times$ 10⁻³ sq. mi.
1 acre	= 0.405 hectares; 1 hectare = 2.471 acres
1 sq. km.	= 0.386 sq. mi.; 1 sq. mi. = 2.590 sq. km.
	= 247.1 acres; 1 acre = 4.047 $\times$ 10⁻³ sq. km.
	= 100 hectares; 1 hectare = 0.010 sq. km.

VOLUME

1 cu. m.	= 10⁶ cu. cm.; 1 cu. cm. = 10⁻⁶ cu. m.
	= 61,024 cu. in.; 1 cu. in. = 1.639 $\times$ 10⁻⁵ cu. m.
	= 35.31 cu. ft.; 1 cu. ft. = 0.028 cu. m.
	= 1.308 cu. yds.; 1 cu. yd. = 0.765 cu. m.
1 cu. cm.	= 0.061 cu. in.; 1 cu. in. = 16.39 cu. cm.
	= 3.531 $\times$ 10⁻⁵ cu. ft.; 1 cu. ft. = 28,317 cu. cm.
1 cu. in.	= 5.787 $\times$ 10⁻⁴ cu. ft.; 1 cu. ft. = 1,728 cu. in.
1 cu. ft.	= 0.037 cu. yd.; 1 cu. yd. = 27 cu. ft.

CAPACITY (LIQUID MEASURE)

1 ltr.	= 103 cu. cm.; 1 cu. cm. = 10⁻³ l.
	= 61.02 cu. in.; 1 cu. in. = 0.164 l.
	= 0.035 cu. ft.; 1 cu. ft. = 28.33 l.
	= 33.81 fl. oz.(U.S.); 1 fl. oz.(U.S.) = 29.56 $\times$ 10⁻³ l.
	= 2.112 pt.; 1 pt. = 0.473 l.
	= 1.056 qt.; 1 qt. = 0.947 l.
	= 0.264 gal.; 1 gal. = 3.788 l.
1 cu. in.	= 0.554 fl. oz.; 1 fl. oz. = 1.805 cu. in.
	= 0.035 pt.; 1 pt. = 28.87 cu. in.
1 cu. ft.	= 29.92 qt.; 1 qt. = 0.033 cu. ft.
	= 7.481 gal.; 1 gal. = 0.134 cu. ft.
1 cup	= 8 fl. oz.; 1 fl. oz. = 0.125 cups
1 fl. oz.	= 0.0625 pt.; 1 pt. = 16 fl. oz.
1 cu. m.	= 264.173 gal.; 1 gal. = 3.785 $\times$ 10⁻³ cu. m.
1 gal.(imp.)	= 1.201 gal.(U.S.); 1 gal.(U.S.) = 0.833 gal.(imp.)

LIQUID FLOW RATE

1 l. per sec. = 15.85 gal. per min.

1 gal. per
min. = 0.063 l. per sec.

FUEL CONSUMPTION

1 ltr. per
100 km. = 0.425 gal. per 100 mi.

1 gal. per
100 mi. = 2.352 l. per 100 km.

MILEAGE

1 km. per l. = 2.353 mi. per gal. (MPG)

1 mi. per
gal. = 0.425 km. per l.

MIXTURE BY VOLUME (LIQUID)

1 ml. per l. = 0.184 fl. oz. per gal.

1 fl. oz. per
gal. = 5.43 ml. per l.

MASS

1 kg. = 1,000 g.; 1 g. = 10^{-3} kg.
= 2.205 lb.; 1 lb. = 0.454 kg.
= 1.102×10^{-3} tons (short); 1 ton (short) = 907.4 kg.
= 10^{-3} tonnes; 1 tonne (metric) = 1,000 kg.

1 g. = 0.035 oz.; 1 oz. = 28.57 g.
= 2.188×10^{-3} lb.; 1 lb. = 457.1 g.

1 oz. = 0.062 lb.; 1 lb. = 16 oz.

1 lb. = 10^{-2} cwt. (short); 1 cwt. (short) = 100 lb.
= 5×10^{-4} tons (short); 1 ton (short) = 2,000 lb.
= 4.54×10^{-4} tonnes; 1 tonne (metric) = 2,205 lb.

1 ton
(short) = 0.907 tonnes; 1 tonne = 1.102 tons (short)

DENSITY

1 kg./cu. m. = 1,000 g./cu. cm.; 1 g./cu. cm. = 10^{-3} kg./cu. cm.
= 0.062 lb./cu. ft.; 1 lb./cu. ft. = 16.02 kg./cu. m.

PRESSURE

1 kg./sq. cm. = 14.22 lb./sq. in.; 1 lb./sq. in. = 0.070 kg./sq. cm.
= 2,048 lb./sq. ft.; 1 lb./sq. ft. = 4.883×10^{-4} kg./sq. cm.
= 98.06 kpa.; 1 kpa. = 0.010 kg./sq. cm.
= 73.55 cm. Hg.; 1 cm. Hg. = 0.013 kg./sq. cm.
= 0.968 atm.; 1 atm. = 1.033 kg./sq. cm.

1 lb./sq. in. = 6.896 kpa.; 1 kpa. = 0.145 lb./sq. in.

$= 5.171$ cm. Hg.; 1 cm. Hg. $= 0.193$ lb./sq. in.

$= 0.068$ atm.; 1 atm. $= 14.7$ lb./sq. in.

1 millibar $= 100$ pa.; 1 pa. $= 0.01$ millibars

VELOCITY

1 km./hr. $= 0.278$ m./sec.; 1 m./sec. $= 3.600$ km./hr.

               $= 0.911$ ft./sec.; 1 ft./sec. $= 1.097$ km./hr.

               $= 0.621$ mi./hr.; 1 mi./hr. $= 1.609$ km./hr.

1 m./sec. $= 3.281$ ft./sec.; 1 ft./sec. $= 0.305$ m./sec.

               $= 2.237$ mi./hr.; 1 mi./hr. $= 0.447$ m./sec.

1 ft./sec. $= 0.682$ mi./hr.; 1 mi./hr. $= 1.467$ ft./sec.

ANGULAR MEASURE

1 radian $= 57.30$ degrees; 1 degree $= 0.174$ radians

Appendix C

More about Money

This appendix amplifies on some of the topics covered in Chapters 2 and 3. First, it discusses two concepts central to most financial calculations: compound interest and the present value of future payments. It also covers the effects of taxation, the effects of different compounding periods, and the effects of inflation on future buying power. Finally, it gives the basis on which each of Tables 2.3 through 2.6 and 3.1 through 3.6 has been calculated.

C.1—Compound Interest

When you put money into a savings account, it earns interest; when you borrow money, you must pay interest. Simple interest is calculated as a flat percentage of the principal (the amount deposited or borrowed); when interest is added to the original amount and then interest is calculated on the combined amount, you enter the world of compound interest.

 If you deposit $100 in a savings account at an annual rate of 5%, the interest you will have earned at the end of one year, will be 5% of $100 = $5. This is added to the original principal—and, in the second year, the interest earned will be 5% of $105 = $5.25, giving a new balance of $110.25. During the third year, the interest earned will be 5% of $110.25 = $5.51—and so on.

Because the interest is added once per year in this example, it is said to be compounded annually. In other cases, it might be added semi-annually, or quarterly, or monthly. Or, as is the usual practice today, it might be added 365 times a year in a daily-interest savings account. (To be precise, the interest in a daily-interest savings account is calculated on a daily basis, but it is credited to the account only at the end of the month. This small difference can be overlooked in the present discussion.)

Because the amount of interest your money earns can be significantly different for different compounding periods, you should be careful to use the right period when you are considering interest rates.

Table C.1 is a general-purpose table of compound-interest factors. The time periods that make up the rows in the table may be years, months, days, or any other period. The columns are for various values of interest rates per period. In most cases, the periods will be years and the rates per period will be annual rates. The factors shown in the table are the future values of $1 at the end of the particular period of time, after interest at the particular rate has been added.

EXAMPLE From Table C.1, you can see that $1 lent, borrowed, or invested for three years at 3% annually will amount to $1.09273 at the end of three years. In the same way, $1 lent for three months at 3% per month also amounts to $1.09273. However, $1 lent for one year at 3% per month amounts to $1.42576, because, in this case, one year represents twelve periods at 3% per period.

Table C.1 is computed in the following way: A principal P invested at a rate of interest r and compounded q times a year for n years will grow to an amount An according to the following formula: $An = P(1 + r \div q)^{nq}$. For example, for $P = \$100$ invested at $r = 10\%$ compounded once a year for ten years, you have: $An = 100(1 + 10 \div 100)^{10} = 100(1.1)^{10}$, and using a calculator to work out the value of $(1.1)^{10}$, you find the value 2.5937. $100 \times 2.5937 = \$259.37$—the amount of interest compounded in this case. You can confirm that this is the value given in Table C.1.

Table C.1 can be used to compare alternative investments or loans in the following way:

EXAMPLE Two alternative investments might be available for your consideration—one offering 20% interest compounded quarterly, and the other offering 21% compounded annually. To find which of the two investments is preferable, consider that for the 20% case there are four compounding periods in a year, so the rate per period is $20 \div 4 = 5\%$. In the 5% column at

four periods, Table C.1 shows the factor
1.21551. Since this is larger than the factor
1.21000 for the 21% annual case, the conclusion is that 20%, compounded quarterly, is a
better choice. Conversely, if you are considering two loans at these rates, the one carrying
21% compounded annually is a better choice.

Although Table C.1 is useful in many calculations, in other cases it
will be better to use special-purpose tables that include the effects of
taxation, factors based on daily-interest rates rather than annual
rates, and the like. Various special tables are included in the
calculations covered in Chapters 2 and 3.

TABLE C.1

Compound Interest Factor Table

Interest Rate Per Period

		1%	2%	3%	4%	5%	6%	7%
	1	1.01000	1.02000	1.03000	1.04000	1.05000	1.0600	1.0700
	2	1.02010	1.04040	1.06090	1.08160	1.10250	1.1236	1.1449
	3	1.03030	1.06121	1.09273	1.12486	1.15762	1.1910	1.2250
	4	1.04060	1.08243	1.12551	1.16986	1.21551	1.2625	1.3108
	5	1.05101	1.10408	1.15927	1.21665	1.27628	1.3382	1.4026
	6	1.06152	1.12616	1.19405	1.26532	1.34010	1.4185	1.5007
	7	1.07214	1.14869	1.22987	1.31593	1.40710	1.5036	1.6058
	8	1.08286	1.17166	1.26677	1.36857	1.47746	1.5938	1.7182
P	9	1.09369	1.19509	1.30477	1.42331	1.55133	1.6895	1.8385
	10	1.10462	1.21899	1.34392	1.48024	1.62889	1.7908	1.9672
E	11	1.11567	1.24337	1.38423	1.53945	1.71034	1.8983	2.1049
	12	1.12683	1.26824	1.42576	1.60103	1.79586	2.0122	2.2522
R	13	1.13809	1.29361	1.46853	1.66507	1.88565	2.1329	2.4098
	14	1.14947	1.31948	1.51259	1.73168	1.97993	2.2609	2.5785
I	15	1.16097	1.34587	1.55797	1.80094	2.07893	2.3966	2.7590
	16	1.17258	1.37279	1.60471	1.87298	2.18287	2.5404	2.9522
O	17	1.18430	1.40024	1.65285	1.94790	2.29202	2.6928	3.1588
	18	1.19615	1.42825	1.70243	2.02582	2.40662	2.8543	3.3799
D	19	1.20811	1.45681	1.75351	2.10685	2.52695	3.0256	3.6165
	20	1.22019	1.48595	1.80611	2.19112	2.65330	3.2071	3.8697
S	21	1.23239	1.51567	1.86029	2.27877	2.78596	3.3996	4.1406
	22	1.24472	1.54598	1.91610	2.36992	2.92526	3.6035	4.4304
	23	1.25716	1.57690	1.97359	2.46472	3.07152	3.8197	4.7405
	24	1.26973	1.60844	2.03279	2.56330	3.22510	4.0489	5.0724
	25	1.28243	1.64061	2.09378	2.66584	3.38635	4.2919	5.4274
	26	1.29526	1.67342	2.15659	2.77247	3.55567	4.5494	5.8074
	27	1.30821	1.70689	2.22129	2.88337	3.73349	4.8223	6.2139
	28	1.32129	1.74102	2.28793	2.99870	3.92013	5.1117	6.6488
	29	1.33450	1.77584	2.35657	3.11865	4.11614	5.4184	7.1143
	30	1.34785	1.81136	2.42726	3.24340	4.32194	5.7435	7.6123
	31	1.36133	1.84759	2.50008	3.37313	4.53804	6.0881	8.1451
	32	1.37494	1.88454	2.57508	3.50806	4.76494	6.4534	8.7153
	33	1.38869	1.92223	2.65234	3.64838	5.00319	6.8406	9.3253
	34	1.40258	1.96068	2.73191	3.79432	5.25335	7.2510	9.9781
	35	1.41660	1.99989	2.81386	3.94609	5.51602	7.6861	10.6766
	36	1.43077	2.03989	2.89828	4.10393	5.79182	8.1473	11.4239

	1%	2%	Interest Rate Per Period 3%	4%	5%	6%	7%
37	1.44508	2.08069	2.98523	4.26809	6.08141	8.6361	12.2236
38	1.45953	2.12230	3.07478	4.43881	6.38548	9.1543	13.0793
39	1.47412	2.16474	3.16703	4.61637	6.70475	9.7035	13.9948
40	1.48886	2.20804	3.26204	4.80102	7.03999	10.2857	14.9745

TABLE C.1—COMPOUND INTEREST FACTOR TABLE (CONTINUED)

		8%	9%	Interest Rate Per Period 10%	11%	12%	13%	14%
	1	1.0800	1.0900	1.1000	1.1100	1.1200	1.1300	1.140
	2	1.1664	1.1881	1.2100	1.2321	1.2544	1.277	1.300
	3	1.2597	1.2950	1.3310	1.3676	1.4049	1.443	1.482
	4	1.3605	1.4116	1.4641	1.5181	1.5735	1.630	1.689
	5	1.4693	1.5386	1.6105	1.6851	1.7623	1.842	1.925
	6	1.5869	1.6771	1.7716	1.8704	1.9738	2.082	2.195
	7	1.7138	1.8280	1.9487	2.0762	2.2107	2.353	2.502
	8	1.8509	1.9926	2.1436	2.3045	2.4760	2.658	2.853
P	9	1.9990	2.1719	2.3579	2.5580	2.7731	3.004	3.252
	10	2.1589	2.3674	2.5937	2.8394	3.1058	3.395	3.707
E	11	2.3316	2.5804	2.8531	3.1518	3.4785	3.836	4.226
	12	2.5182	2.8127	3.1384	3.4985	3.8960	4.335	4.818
R	13	2.7196	3.0658	3.4523	3.8833	4.3635	4.898	5.492
	14	2.9372	3.3417	3.7975	4.3104	4.8871	5.535	6.261
I	15	3.1722	3.6425	4.1772	4.7846	5.4736	6.254	7.138
	16	3.4259	3.9703	4.5950	5.3109	6.1304	7.067	8.137
O	17	3.7000	4.3276	5.0545	5.8951	6.8660	7.986	9.276
	18	3.9960	4.7171	5.5599	6.5436	7.6900	9.024	10.575
D	19	4.3157	5.1417	6.1159	7.2633	8.6128	10.197	12.056
	20	4.6610	5.6044	6.7275	8.0623	9.6463	11.523	13.744
S	21	5.0338	6.1088	7.4002	8.9492	10.8038	13.021	15.668
	22	5.4365	6.6586	8.1403	9.9336	12.1003	14.714	17.861
	23	5.8715	7.2579	8.9543	11.0263	13.5523	16.627	20.362
	24	6.3412	7.9111	9.8497	12.2382	15.1786	18.788	23.212
	25	6.8485	8.6231	10.8347	13.5855	17.0001	21.231	26.462
	26	7.3964	9.3992	11.9182	15.0799	19.0401	23.991	30.167
	27	7.9881	10.2451	13.1100	16.7386	21.3249	27.109	34.390
	28	8.6271	11.1671	14.4210	18.5799	23.8839	30.633	39.204
	29	9.3173	12.1722	15.8631	20.6237	26.7499	34.616	44.693
	30	10.0627	13.2677	17.4494	22.8923	29.9599	39.116	50.950
	31	10.8677	14.4618	19.1943	25.4104	33.5551	44.201	58.083
	32	11.7371	15.7633	21.1138	28.2056	37.5817	49.947	66.215
	33	12.6760	17.1820	23.2252	31.3082	42.0915	56.440	75.485
	34	13.6901	18.7284	25.5477	34.7521	47.1425	63.777	86.053
	35	14.7853	20.4140	28.1024	38.5749	52.7996	72.069	98.100
	36	15.9682	22.2512	30.9127	42.8181	59.1356	81.437	111.834

	8%	9%	Interest Rate Per Period 10%	11%	12%	13%	14%
37	17.2456	24.2538	34.0039	47.5281	66.2318	92.024	127.491
38	18.6253	26.4367	37.4043	52.7562	74.1797	103.987	145.340
39	20.1153	28.8160	41.1448	58.5593	83.0182	117.506	165.687
40	21.7245	31.4094	45.2593	65.0009	93.0510	132.782	188.884

TABLE C.1—COMPOUND INTEREST FACTOR TABLE (CONTINUED)

| | | Interest Rate Per Period | | | | | |
		15%	16%	17%	18%	19%	20%
	1	1.150	1.160	1.170	1.180	1.19	1.20
	2	1.322	1.346	1.369	1.392	1.42	1.44
	3	1.521	1.561	1.602	1.643	1.69	1.73
	4	1.749	1.811	1.874	1.930	2.01	2.07
	5	2.011	2.100	2.192	2.288	2.39	2.49
	6	2.313	2.436	2.565	2.700	2.84	2.99
	7	2.660	2.826	3.001	3.185	3.38	3.58
	8	3.059	3.278	3.511	3.759	4.02	4.30
	9	3.518	3.803	4.108	4.435	4.79	5.16
P	10	4.046	4.411	4.807	5.234	5.69	6.19
	11	4.652	5.117	5.624	6.176	6.78	7.43
E	12	5.350	5.936	6.580	7.288	8.06	8.92
	13	6.153	6.886	7.699	8.599	9.60	10.70
R	14	7.076	7.988	9.007	10.147	11.42	12.84
	15	8.137	9.266	10.539	11.974	13.59	15.41
I	16	9.358	10.748	12.330	14.129	16.17	18.49
	17	10.761	12.468	14.426	16.672	19.24	22.19
O	18	12.375	14.463	16.879	19.673	22.90	26.62
	19	14.232	16.777	19.748	23.214	27.25	31.95
D	20	16.367	19.461	23.106	27.393	32.43	38.34
	21	18.822	22.574	27.034	32.324	38.59	46.01
S	22	21.645	26.186	31.629	38.142	45.92	55.21
	23	24.891	30.376	37.006	45.008	54.65	66.25
	24	28.625	35.236	43.207	53.109	65.03	79.50
	25	32.919	40.874	50.658	62.669	77.39	95.40
	26	37.857	47.414	59.270	73.949	92.09	114.48
	27	43.535	55.000	69.345	87.260	109.59	137.37
	28	50.066	63.800	81.134	102.967	130.41	164.84
	29	57.575	74.009	94.927	121.501	155.19	197.81
	30	66.212	85.950	111.065	143.371	184.68	237.38
	31	76.144	99.586	129.946	169.177	219.76	284.85
	32	87.565	115.520	152.036	199.629	261.52	341.82
	33	100.700	134.003	177.883	235.563	311.21	410.19
	34	115.805	155.443	208.123	277.964	370.34	492.22
	35	133.176	180.314	243.503	327.997	440.70	590.67
	36	153.152	209.164	284.899	387.037	524.43	708.80

	15%	16%	Interest Rate Per Period 17%	18%	19%	20%
37	176.125	242.631	333.332	456.703	624.08	850.56
38	202.543	281.452	389.998	538.910	742.65	1020.67
39	232.925	326.484	456.298	635.914	883.75	1224.81
40	267.864	378.721	533.869	750.378	1051.67	1469.77

C.2—What Present Value Means

Money that you possess today can be used immediately—either by spending it or by investing it so that it begins earning interest. On the other hand, money that is owed, either to you or by you, at some time in the future, has a value different from the same money's value today, and its value is related in some way to how far in the future it is due. This concept brings in the time value of money.

Using a common definition, the present value of an amount of money due in the future is the amount that you would need to invest today, at a certain rate of interest, to have it grow to the same value at the same time in the future.

 If the rate of interest that you can earn is 5%, compounded annually, then the present value of $100 due one year in the future is about $95, since that amount invested at 5% will grow to about $100 in one year.

Note that present value changes if the rate of interest changes.

 If you can invest money at 10% instead of 5%, the present value of $100 due one year in the future will be about $91 instead of $95, since $100 = $91 plus 10% of $91.

The method of calculating the present value of a sum is similar to the method of calculating compound interest outlined in C.1, page 307. The formula in this case is: $P = Anq (1 + r \div q)^{nq}$, where P is the present value, A is the amount, n is the amount of time the interest calculated, q is the number of times per year interest is compounded, and r is the interest rate. Table 3.2 is calculated in this way in Chapter 3.

Different amounts owed at different times in the future can be added together by expressing each one in terms of its present value. This is particularly useful when considering a series of payments to be made at regular intervals in the future, such as the proceeds from an insurance policy. Such regular payments are called annuities.

 A series of three deposits, each of $100, are to be made at yearly intervals, starting one year from now, which will earn interest at 6% per year. So:

- The first deposit is made at the end of the first year, and it will remain on deposit for two years— earning $6 interest the first year and $6.36 the next, for a total of $12.36.

- The second deposit is made at the end of the second year, and it will remain on deposit for one year— earning $6 interest.
- The third deposit is made at the end of the third year, and earns no interest since that is also the end of the period.
- The total interest earned will be $18.36, and the balance at the end of the period will be $318.36. This is called the future value of the series of deposits.

The present value of the series in this case is the single sum that you would need to deposit at a certain interest rate that would amount to $318.36 in three years.

Assuming that you can earn 6%, you can find the present value using Table 3.2. The entry there is $84 for the present value of $100, and by multiplying this by 318.36 ÷ 100, you obtain $267.42.

The concept of present value is important, since it makes it possible to compare future earnings or future expenses on a common basis. One case where it is particularly useful is in deciding whether you should make a certain investment. You should calculate the present value of the investment—and if the result is greater than the present value of the money to be invested, you can conclude that it is worthwhile making the investment. Examples of this type of calculation are included in Chapters 2 and 3.

C.3—Effect of Taxation

For many people, the interest they earn on savings is subject to tax—whether or not it is withdrawn for use as income. For this reason, it is important to consider the effects of taxation whenever you are making calculations concerning the interest earned on savings. In most cases, net tax rates of 15%, 28%, or 31% will be applicable, because these are the current rates of federal taxes. If necessary, the results can be adjusted for other rates.

In this context, the term "marginal tax rate" is frequently used because interest on savings is usually income that is added to other sources. This means that the basic tax exemptions have already been applied and each additional dollar will be taxed at the highest rate that will apply. The marginal tax rate is the rate that applies to the last dollar earned.

C.4—Different Compounding Periods

Several of the tables in Chapters 2 and 3 are drawn up for cases in which interest is compounded daily. In order to find results for annual, semi-annual, quarterly, or monthly compounding, you need to make some adjustments. This is done by finding an equivalent daily interest rate using Table C.4.

TABLE C.4

Daily Interest Rates Equivalent to Annual, Semi-annual, Quarterly and Monthly Rates

CONVERTING ANNUAL TO DAILY		CONVERTING SEMI-ANNUAL TO DAILY		CONVERTING QUARTERLY TO DAILY		CONVERTING MONTHLY TO DAILY	
ANNU-AL RATE	EQUIV. DAILY RATE	SEMI-ANNU-AL RATE	EQUIV. DAILY RATE	QUART-ERLY RATE	EQUIV. DAILY RATE	MONTH-LY RATE	EQUIV. DAILY RATE
4%	3.92	4%	3.96	4%	3.98	4%	4.00
5%	4.88	5%	4.94	5%	4.97	5%	4.99
6%	5.83	6%	5.91	6%	5.96	6%	5.98
7%	6.77	7%	6.88	7%	6.94	7%	6.98
8%	7.70	8%	7.85	8%	7.93	8%	7.97
9%	8.62	9%	8.81	9%	8.90	9%	8.97
10%	9.53	10%	9.76	10%	9.88	10%	9.96
11%	10.44	11%	10.71	11%	10.86	11%	10.95
12%	11.33	12%	11.66	12%	11.83	12%	11.94
13%	12.23	13%	12.61	13%	12.81	13%	12.94
14%	13.11	14%	13.55	14%	13.78	14%	13.93
15%	13.98	15%	14.48	15%	14.74	15%	14.92

This table shows, for instance, that if you want to find the interest earned in an account in which the interest rate is 4% compounded annually, the same answer will be obtained by finding an equivalent daily interest rate of 3.92% from Table C.4 and entering Table 3.1 with this rate.

To find the interest earned after ten years on $100 at an interest rate of 9% compounded quarterly (assuming you don't have to pay taxes), Table C.4 shows that the equivalent daily interest rate is 8.9%. From Table 3.1, the amount for 8% is $122, and the amount for 9% is $146. By the interpolation formula in A.2, page 247, the amount for 8.9% is: 122 + 0.9 × (146 − 122) = $144.

C.5—Effects of Inflation on Future Buying Power

Experience has shown that the buying power of money is likely to decline as time goes by. The Consumer Price Index and its variation through the years are discussed in Section 1.5, page 15. We illustrate there how buying power changes from year to year. Although you can't confidently predict the inflation rate, you should always include some estimate when you are making calculations involving the future value of money. Many of the calculations in this appendix are concerned with future values—and, therefore, the results should take account of the probable effects of inflation. In Table C.5, the relative buying power of $100 now and at future dates is shown for various assumed inflation rates:

TABLE C.5

Future Buying Power Relative to $100 Today
(Dollars)

YEARS AHEAD	CONSTANT ANNUAL INFLATION RATE ASSUMED											
	1%	2%	3%	4%	5%	6%	7%	8%	9%	10%	12%	15%
1	99	98	97	96	95	94	93	93	92	91	89	87
2	98	96	94	92	91	89	87	86	84	83	80	76
3	97	94	92	89	86	84	82	79	77	75	71	66
4	96	92	89	85	82	79	76	74	71	68	64	57
5	95	91	86	82	78	75	71	68	65	62	57	50
6	94	89	84	79	75	70	67	63	60	56	51	43
7	93	87	81	76	71	67	62	58	55	51	45	38
8	92	85	79	73	68	63	58	54	50	47	40	33
9	91	84	77	70	64	59	54	50	46	42	36	28
10	91	82	74	68	61	56	51	46	42	39	32	25
12	89	79	70	62	56	50	44	40	36	32	26	19
15	86	74	64	56	48	42	36	32	27	24	18	12

C.6—Mathematical Basis for Tables 2.3 – 2.6 and 3.1 – 3.6

⬅ Mathematical Basis for Table 2.3

Table 2.3 has the same mathematical basis as Table 2.4, which in turn has the same basis as Table 3.5 except for one minor change. (See the discussion for Table 2.4.)

⬅ Mathematical Basis for Table 2.4

The monthly payments on a mortgage are calculated in the same way that the present value of a series of monthly payments is calculated. This means that the mathematical basis of Table 2.4 is the same as that of Table 3.5. (See C.6, page 325.) The only difference is that the compounding period for the mortgage calculations is semi-annual while Table 3.5 uses daily compounding.

For example, Table 3.5 shows that the present value of a series of equal monthly payments of $1 at 10% daily interest payable over five years is $47. The equivalent daily interest from Table C.4, page 319, would be 9.76%, so the present value of the series would be $47.25. If the mortgage amount is $1,000, the present value required is $1,000—and, therefore, the monthly payment needed is $1,000 ÷ 47.25 = $21.16, and this is the amount shown in Table 2.4 for this case.

⬅ Mathematical Basis for Table 2.5

The upper portion of the table corresponds to Table 3.1, except that the balance in the account is shown instead of the interest earned. Consequently, it has the same mathematical basis. (See C.6, page 323.

The lower portions are calculated in the same way, except that the initial principal is reduced by the amount of the tax payable and the interest rate is reduced by the tax rate, as discussed for Table 3.2.

✏ Mathematical Basis for Table 2.6

As explained in Section 2.6, page 37, entries in Table 2.6 correspond to those in Table 3.5—and, therefore, this table has the same mathematical basis as Table 3.5.

✏ Mathematical Basis for Table 3.1

The mathematical basis on which Table 3.1 has been computed is similar to that for Table C.1, and the same formula is used. For example, for $P = \$100$ invested at $r = 8\%$ compounded daily for 10 years, the formula in C.1, page 308 gives: $An = 100 (1 + 0.08 \div 365)^{10 \times 365}$.

Using a calculator to work out the answer gives the result: $An = 222$—and, therefore, the interest earned is $An - P = 222 - 100 = 122$. To confirm the result, you can check Table 3.1 for this case and find the entry is 122.

The same example can be extended to illustrate the effect of taxes, which is to reduce the interest rate by the percentage of tax deducted. For example, if 28% taxes are payable, the effective interest rate is: $8\% - (28\%$ of $8\%) = 5.76\%$, and when this is substituted in the equation, it becomes: $An = 100 (1 + 0.0576 \div 365)^{10 \times 365}$.

Using a calculator to work out the answer gives the result: $An = 179$—and, therefore, the interest is $179 - 100 = 79$.

To confirm this result, you can check Table 3.1 for this case and find the entry 79.

✏ Mathematical Basis for Table 3.2

By the definition of present value given in C.2, page 316, it is evident that Table 3.2 also shows the present value of a future payment because it gives the amount that must be invested today at a certain rate of interest to grow to the same amount at the same time in the future. This means that the formula given on page 316 for calculating the present value of a sum provides the mathematical basis for Table 3.2.

For example, the deposit needed to accumulate $100 over five years with 6% daily interest can be found using the formula on page 316: $P = 100 (1 + 0.06 \div 365)^{-5 \times 365}$.

Using a calculator to evaluate P for this case gives the result $74. To confirm this result, you can check Table 3.2 for this case and find the entry 74 when no tax is payable. If 28% tax is payable, the effective interest rate becomes 6%—(28% of 6%) = 4.3%. In this case, $P =$ $80 as shown in Table 3.2—and this amount can be checked by substituting in the formula and using a calculator.

Mathematical Basis for Table 3.3

Objective: to express the future value of a series of monthly deposits in an account paying daily interest. The formula for this is: $F = d \div R[(1 + R)^{12y} - 1]$, where F is the future value, d is the monthly deposit, y is the number of years, and R is the effective interest rate, which is related to the given interest rate r by the formula: $R = (1 + r \div 365)^{365 \div 12} - 1$.

For example, if the interest rate r is 9%, $R = (1 + 0.09 \div 365)^{365 \div 12} - 1$.

Using a calculator to solve for R gives the value 0.00753. Now if, for example, $1 is deposited monthly for a period of ten years, the future value $F = 1 \div 0.00753[(1.00753)^{120} - 1]$.

Using a calculator to solve for F gives the value 193, and subtracting the amount of the 120 deposits made over the ten-year period gives the value $73 for the interest earned. This is the value shown for this case in Table 3.3.

To find the answer when taxes are payable, the interest rate r should be reduced by the tax rate before solving the equations. (See the previous discussion concerning Table 3.2.)

✏️ Mathematical Basis for Table 3.4

Objective: to find the monthly deposit that must be made to reach a known future value. The formula used to calculate Table 3.3 can be used again, except that this time the value of F is known and it is required to solve the equation for the value of d, the monthly deposit.

For example, as before, if the interest rate is 9%, then $R = 0.00753$. If the desired future value is $1,000 and $n = 10$ years, the equation becomes: $1000 = d \div R [(1 + R)^{120} - 1]$.

Using a calculator to solve the equation for d gives the value $d = 5.2$, which is the value given in Table 3.4.

As for Tables 3.2 and 3.3, to find the answer when taxes are payable, you must reduce the interest rate by the tax rate before solving the equations.

✏️ Mathematical Basis for Table 3.5

Objective: to find the present value of a series of monthly deposits in a daily-interest account. The formula that applies is: $P = d \div R [1 - (1 + R)^{-12y}$ where P is the present value and the other symbols are as defined for Table 3.3. For example, if $r = 10\%$, the formula for R gives: $R = (1 + 0.10 \div 365)^{365 \div 12} - 1 = 0.00837$, and for $d = \$1$ and $y = 5$ years, the formula for P gives: $P = 1 \div 0.00837[1 - (1 + 0.00837)]^{-60}$.

Using a calculator to solve for P gives the value $47, which is the value given for this case in Table 3.5. To find the answer when taxes are payable, you must reduce the interest by the tax rate before making the calculations, as explained for Table 3.2.

✏️ Mathematical Basis for Table 3.6

In this case, if the account balance is to be maintained indefinitely, the amount that can be withdrawn each month is the amount of interest added in that month. As for Table 3.3, the effective interest rate R is given by the formula: $R = (1 + r \div c)^{c \div 12} - 1$.

For example if $r = 10\%$ and $c = 365$, using a calculator to solve the equation for R gives the result 0.00837. For an account of $1,000, the monthly interest added to the account is $1,000 \times R = \$8.37$, which is the value (rounded to $8.40) given for this case in Table 3.6.

To illustrate the way the other entries in the table are calculated, an example can be worked through in detail:

EXAMPLE

You have $1,000 in a daily-interest account earning 10%, and you desire to withdraw money at the rate of $21 per month.

After one month, the interest earned will be $1,000R = \$8.37$. At that time, a withdrawal of $21 will consist of $8.37 interest plus $21 − $8.37 = $12.63 of principal, leaving a new principal of $987.37. During the second month, the interest earned will be $987.37 \times R = \$8.26$. At that time, a withdrawal of $21 will consist of $8.26 interest and $12.74 principal, leaving a new principal of $979.11. This process is continued until the account is exhausted, which in this case occurs after 60 months.

If tax is payable, the process is similar except that the interest rate is reduced by the tax rate, as discussed for Table 3.2.

Suggestions for Further Reading

1 ◆ Consuming Passion

Bureau of Labor Statistics. *Handbook of methods, vol. 2. The consumer price index*, Washington, D.C.: U.S. Department of Labor, 1984.

Consumer Reports. *Buying guide issue*. Mount Vernon, N.Y.: Consumers Union of the U.S., Inc., every December.

Gillis, Jack. *The car book*. New York: Harper & Row, annual.

Meyer, Robert. *Consumer and business mathematics*. New York: Arco, 1975.

2 ◆ Your Money or Your Life

Hummel, Paul M. *Mathematics of finance*. New York: McGraw-Hill, 1971.

Rauche, Nelda W. *Business mathematics*. Englewood Cliffs, N.J.: Prentice-Hall, 1978.

Trost, Stan. *Financial cookbook (computer diskette and documentation)*. San Mateo, Calif.: Electronic Arts, 1984.

Walkers, Art. D. *Dollars and sense*. New York: New York Book Co., 1973.

3 ◆ Bank on It

Campbell, Colin D., and Rosemary G. *An introduction to money and banking*. Hinsdale, Ill.: The Dryden Press, 1975.

Fundamental facts about U.S. money. Atlanta: Federal Reserve Bank of Atlanta, 1986.

Lewis, Michael. *Liar's poker*. New York: W.W. Norton & Co., 1989.

Milner, Wendy L. *Personal money management with your micro*. Blue Ridge Summit, Pa.: Tab Books, 1984.

4 ◆ Health and Fitness Figures

Adams, Catherine F. *Handbook of the nutritional value of foods: in common units*. New York: Dover Publications, 1975.

Cooper, Robert K. *Health and fitness excellence: The scientific action plan*. Boston: Houghton Mifflin, 1989.

Deutsch, Ronald M. *The family guide to better food and better health*. Des Moines: Bantam Books, 1973.

Editors of Prevention Magazine. *Life span-plus*. Emmaus, Pa.: Rodale Press, 1990.

Food and Nutrition Board. *Recommended dietary allowances, 9th revised ed*. Washington, D.C.: U.S. Department of Agriculture, 1975.

Hoffman, Mark, ed. *World almanac and book of facts*. New York: Pharos Books, annual.

Lamb, Laurence E. *What you need to know about food and cooking for health*. New York: The Viking Press, 1973.

National Board of the Young Men's Christian Associations. *The official YMCA fitness programs*. New York: Rawson Associates, 1984.

Thommen, George S. *Is this your day?* New York: Crown Publishing, Inc., 1973.

5 ◆ Weather and the Environment

Boyle, S., and Ardell, J. *The greenhouse effect*. Falmouth, Cornwall, U.K.: New English Library, Hodder and Stoughton, 1989.

Campbell, Tim. *Do-it-yourself weather book*. Birmingham, Ala.: Oxmoor House, 1979.

Commoner, Barry. *Making peace with the planet*. New York: Pantheon Books, 1990.

Douglas, Mark. *Ozone layer peril*. Littleton, Colo.: Pye Publications, 1986.

Elsom, Derek. *Atmospheric pollution: causes, effects and control policies*. New York: B. Blackwell, 1987.

Goudie, Andrew. *The human impact: man's role in environmental change*. Oxford, U.K.: B. Blackwell, 1981.

Hazardous materials, hazardous waste: local management options. Washington, D.C.: International City Management Association, 1987.

Roan, Sharon. *Ozone crisis: the 15-year evaluation of a sudden global emergency*. New York: Wiley, 1989.

Schneider, Stephen. *Global warming: are we entering the greenhouse century?* San Francisco: Sierra Club Books, 1989.

Schumacher, Aileen. *A guide to hazardous materials management: physical characteristics, federal regulations, and response alternatives*. New York: Quorum Books, 1988.

Standard handbook of hazardous waste treatment and disposal. New York: McGraw-Hill, 1989.

6 ◆ Gambling, Cards and Games—Odds and Probabilities

Anderson, Harry. *Harry Anderson's games you can't lose: A general guide for suckers*. New York: Pocket Books, 1989.

Frey, Richard. *According to Hoyle: Rules of games*. Greenwich, Conn.: Fawcett Pubs, 1985.

———. *The official encyclopedia of bridge*. New York: Crown Publishing Co., 1976.

Goren, Charles. *Goren's complete new bridge*. Garden City, N.Y.: Doubleday, 1985.

Knebel, Fletcher. *Poker game, 1st edition*. Garden City, N.Y.: Doubleday, 1983.

Morehead, Albert, ed. *Official rules of card games*. New York: Fawcett Crest, 1986.

Reese, Terence. *Bridge for ambitious players*. London: Gollancz, 1989.

Scarne, John. *Scarne's encyclopedia of games*. New York, Harper & Row, 1973.

————— . *Scarne's guide to casino gambling*. New York: Simon and Schuster, 1978.

Siskin, Bernard. *What are the chances?: Risks, odds and likelihood in everyday life*. New York: Crown Publishing Co., 1989.

Stone, Joe. *The master's book of pool and billiards*. New York: Crown Publishing Co., 1979.

7 ◆ Sports—Scoring and Statistics

Ballisimo, Lou. *The bowler's manual*. Englewood Cliffs, N.J.: Prentice-Hall, 1975.

Boehm, David A., ed. *Guinness sports record book, 1990-91*. New York: Sterling, 1990.

Considine, Tim. *The language of sport*. New York: World Almanac Publications, 1982.

Menke, Frank G. *The encyclopedia of sports*. South Brunswick, N.J.: A.S. Barnes and Co. Inc., 1975.

Reichler, Joseph L., ed. *The baseball encyclopedia*. New York: Macmillan, 1988.

Richards, Jack W., and Hill, Danny. *Complete handbook of sports scoring and record keeping*. West Nyack, N.Y.: Parke Publishing Co., 1974.

Ritter, Lawrence S. *Complete handbook of pro football*. New York: American Library, 1982.

The rule book. New York: St. Martin's, 1987.

Scott, John W. *Step by step basketball fundamentals for the player and coach*. Englewood Cliffs, N.J.: Prentice-Hall, 1989.

Watson, Tom, and Hannigan, Frank. *The rules of golf*. New York: Times Books, 1988.

8 ◆ Home, Hobbies, and Workshop Numbers

Busha, Ed, et al. *The book of heat*. New York: The Stephen Greene Press, 1982.

Forest Products Laboratory. *Wood handbook: Wood as an engineering material.* Washington, D.C.: U.S. Department of Agriculture, 1974.

Moore, Wynn. *Keeping it on the road.* New York: Quill, 1982.

Motor gasolines. Richmond, Calif.: Chevron Research Co., 1985.

Paul, Henry E. *Binoculars and all-purpose telescopes.* Radnor, Pa.: Chilton Book Co., 1965.

Scharff, Robert. *Mathematics for construction, workshop and home.* New York: Popular Science Books, Harper & Row, 1981.

Swezey, Kenneth M. *Formulas, methods, tips and data for home and workshop.* Albany, N.Y.: Popular Science Books, Harper & Row, 1969.

Vivian, John. *Wood heat.* Emmaus, Pa.: Rodale Press, 1976.

9 ◆ Popular Science Calculations

Asimov, Isaac. *The clock we live on.* New York: Abelard, Schuman, 1965.

Bishop, Owen. *Yard sticks of the universe.* New York: Pete Bedrick Books, 1984.

Burns, Marilyn. *This book is about time.* Boston: Little, Brown and Co., 1978.

Goudsmit, Samuel A., and Claiborne, Robert. *Time.* Alexandria, Va.: Time-Life Books, 1980.

McGraw-Hill. *Encyclopedia of science and technology, 6th ed.* New York: McGraw-Hill, 1987.

Serway, R.A. *Physics for scientists and engineers.* Philadelphia: Saunders College Publishing, Harcourt Brace, 1990.

Appendix A ◆ Tools You Might Need

The broad scope of this appendix would require many suggestions for further reading to cover all subjects individually. Instead, if you wish to review a particular area of mathematics, you should find an introductory textbook and then proceed to more advanced levels. Check your public library for such texts.

On the other hand, if you wish to obtain an introductory book covering several of the same areas as this appendix, you could consult one or more of the following:

Blocksma, Mary. *Reading the numbers.* New York: Penguin Books, 1989.

Complete guide to managing your money. Mount Vernon, N.Y.: Consumers Union, 1989.

Donoghue, William E. *Lifetime financial planner.* New York: Harper and Row, 1987.

Hogben, Lancelot. *Mathematics for the million.* New York: W.W. Norton, 1983.

Parson, Russell D. *Essentials of mathematics.* New York: Wiley, 1989.

Sobel, M., and Maltsky, E. *Essentials of mathematics wit, consumer applications.* Lexington, Mass.: Gunn and Co., 1977.

Washington, Allyn J. *Basic technical mathematics, 4th. ed.* Menlo Park, Calif.: Benjamin/Cummings Publishing Co., 1985.

Appendix B ◆ Getting the Units of Measurement Right

Each edition of *The World Almanac and Book of Facts* , published annually by World Almanac in New York, contains a brief summary of U.S. customary units and international metric units, with conversion factors, in the section on "Standard Weights and Measures." The data for the section are supplied by the National Institute of Standards and Technology of the U.S. Commerce Department. Another excellent table of conversion factors is included in *Great International Maths on Keys Book*, published in 1976 by the Texas Instruments Learning Center.

Index

C

c (velocity of light), 218
Calcium, need for and sources of, 85–86
Calculators
  keys for antilogarithms, 252
  keys for logarithms, 251–252
  keys for sine, cosine, and tangent, 268
  use of change-sign key, 249
  use of inverse key, 249
  use of percent key, 10–12
  use of y^x key, 249, 257
Calendars, 219–236
Calories,
  definitions of, 83, 292
  proportions in a balanced diet, 83–84
  number used in exercising, 78–79, 84
Camera settings, 198–199
Cancer,
  possible risks from electromagnetic fields, 118
  treatment with radioactive materials, 237
Capacity (liquid measure), units of, 301
Carat, 292
Carbohydrates, proportion in a healthy diet, 83
Carbon dioxide,
  in air pollution, 113
  levels affecting global warming, 122
Carbon monoxide in air pollution, 108–109
Cardiovascular,
  fitness, 76, 78
  symptoms due to air pollution, 109

Cards, gambling, and games, 126–147
Carpentry, 191
Carpet tiles, 193
Cars,
  cost of buying and owning, 6–9
  Crash Test Ratings, 207
  engine displacements and powers, 204
  exhaust and air pollution, 108, 109–110
  facts and figures, 204–207
  mileage and fuel consumption, 204–205
  octane rating, 205
  oil grades, 205–206
  tire types, sizes, and pressures, 206–207
  trade-in values, 6–9
Celestial magnitude, 201
Celsius scale of temperature, 102, 292, 297
Cement (*see* Concrete)
Centigrade (*see* Celsius)
Centimeters,
  cubic, 296
  units of area, 296
  units of length, 296, 300
  of mercury, units of pressure, 302, 303
Central Tendency, measures of, 275–280
Ceramic tiles, 193
CFCs (chlorofluorocarbons), 107
Chance,
  games of, 127
  of winning lotteries, 132–133
Chaos, science of, 101
Chlorine, effect on ozone layer, 107

P

GO FIGURE!

GO FIGURE!

GO FIGURE!

GO FIGURE!

GO FIGURE!